HARCOURT Math

INTERVENTION
Problem Solving

Grade 3

Harcourt

Orlando   Austin   Chicago   New York   Toronto   London   San Diego

Visit *The Learning Site!*
www.harcourtschool.com

CONTENTS

Using Math Maps

Use an Intervention • Problem Solving Math Map to

- see how the parts of a word problem fit together.

- show what you know and what you are to find.

- write a number sentence or equation to represent a problem.

- solve a problem.

Marcus has 13 stickers, and Juan has 6 stickers. How many more stickers does Marcus have than Juan?

Larger Amount
13 stickers

Smaller Amount
6 stickers

Difference
?

Join

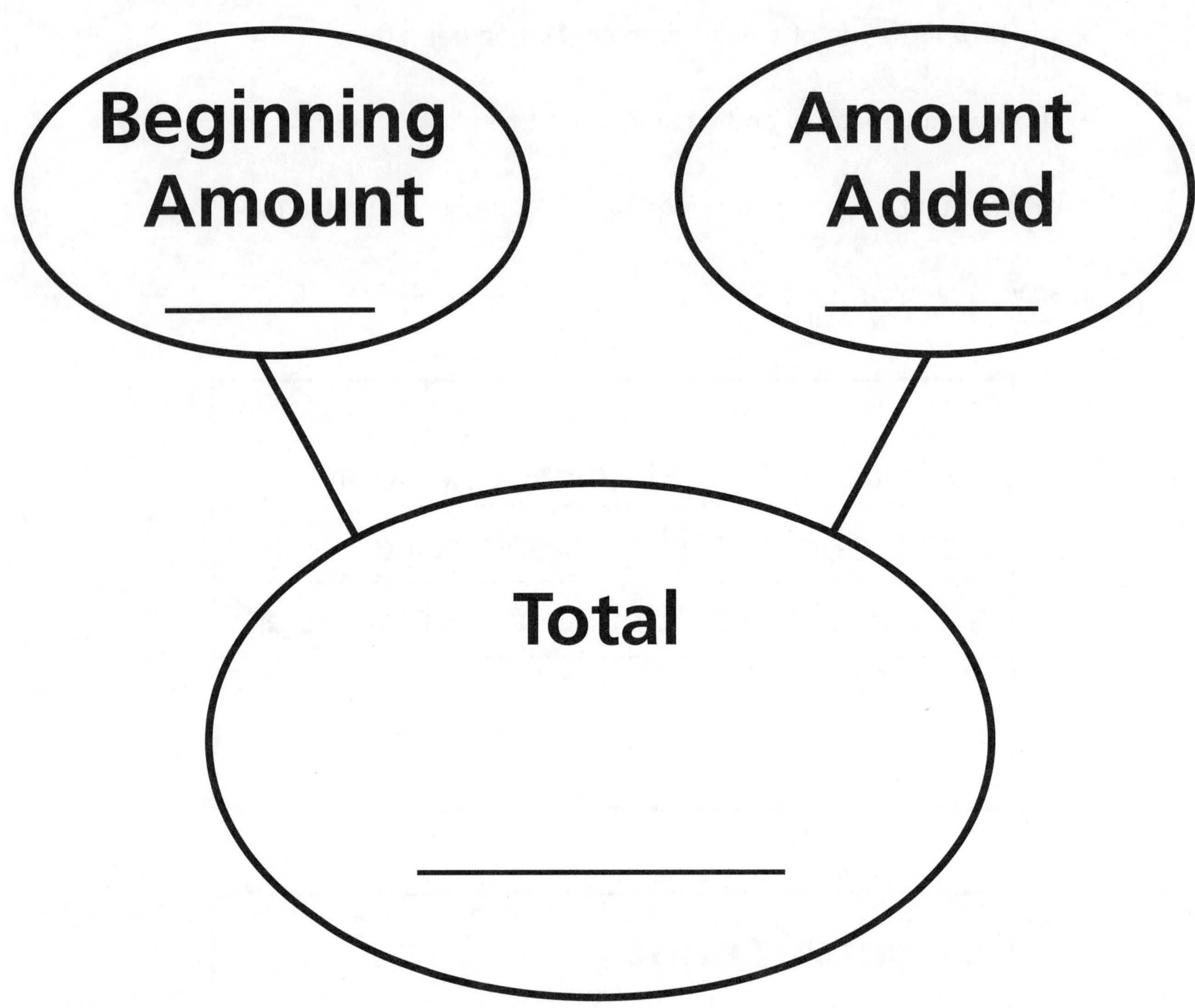

Add or Subtract

Separate

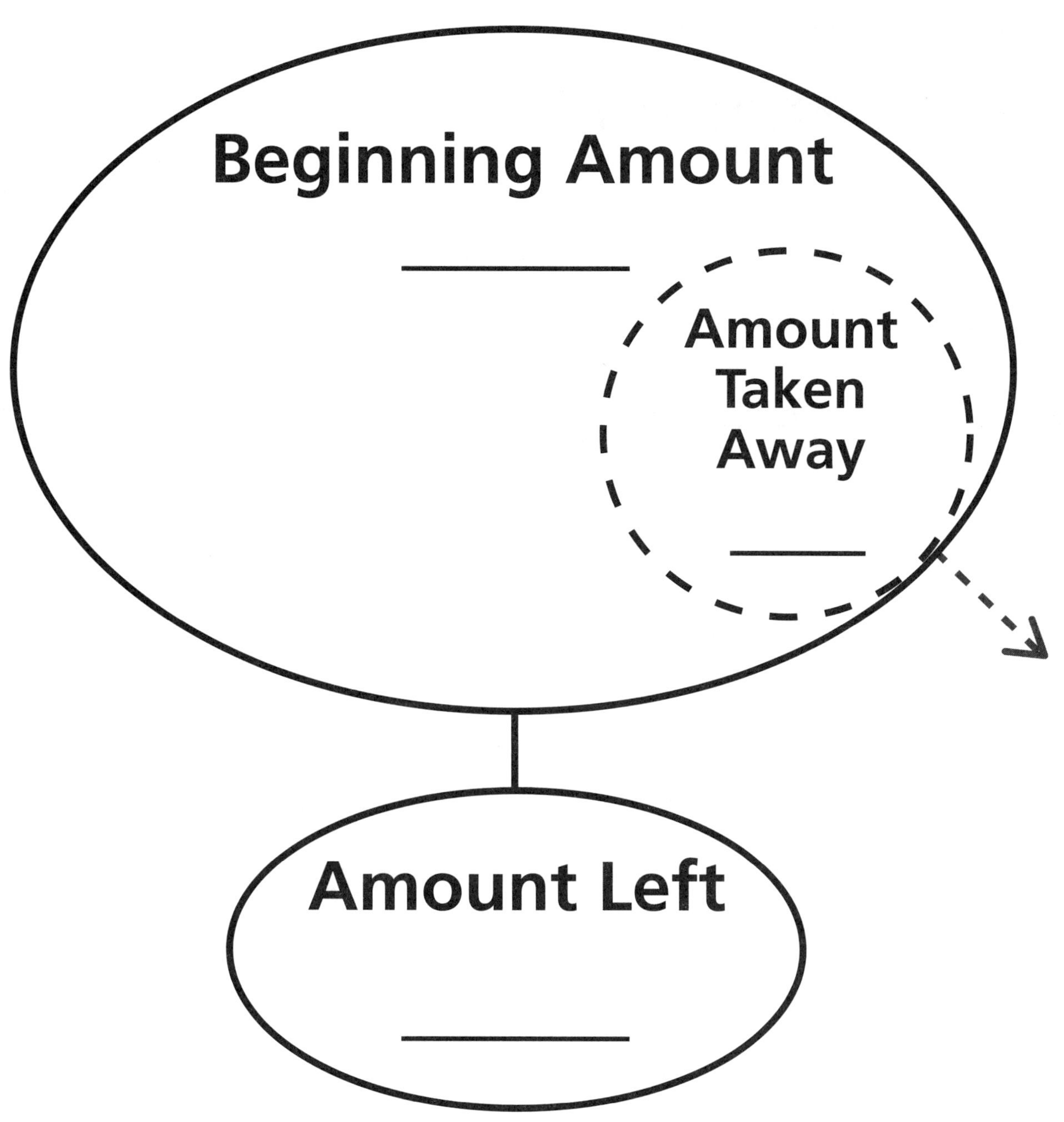

Add or Subtract

Compare

Larger Amount

____

Smaller Amount   **Difference**

_______   _______

Add or Subtract

Part-Part-Whole

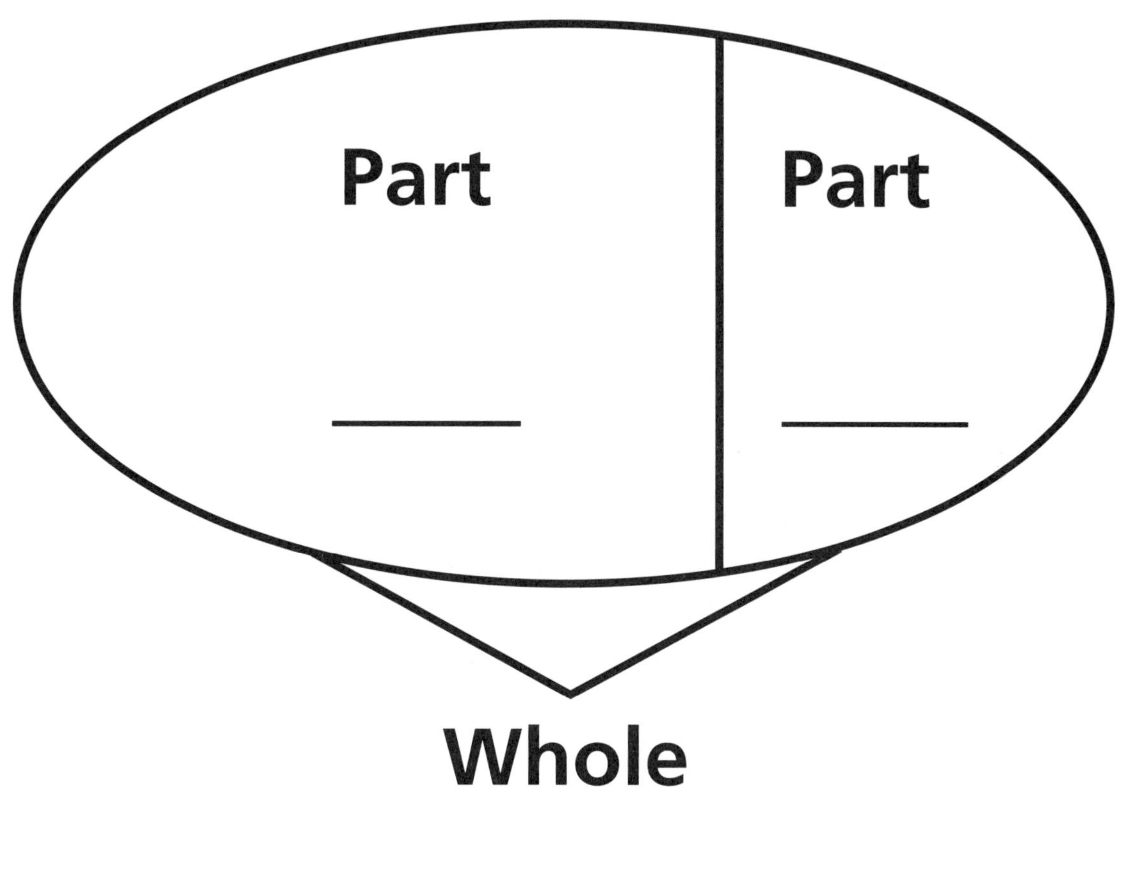

Add or Subtract

Equal Groups

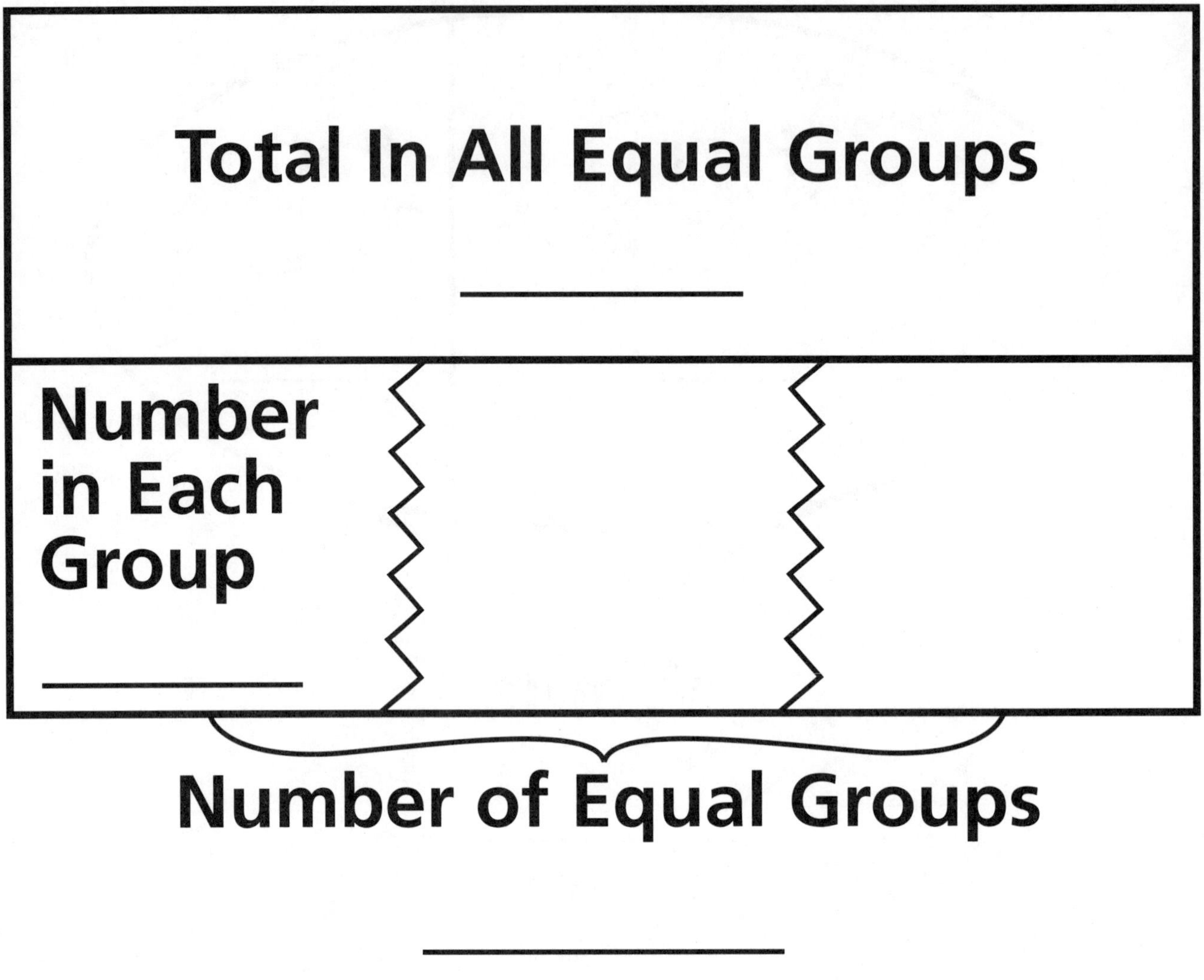

Multiply or Divide

Problem Solving

Understand

1. Retell the problem in your own words. ______________________

____

____

2. List the information given. ______________________________

____

3. Restate the question as a fill-in-the-blank sentence.

____

Plan

4. List one or more problem-solving strategies that you can use.

____

5. Predict what your answer will be. ______________________

____

Solve

6. Show how you solved the problem. ______________________

____

____

7. Write your answer in a complete sentence. ______________

____

Check

8. Tell how you know your answer is reasonable. ____________

____

9. Describe another way you could have solved the problem.

____

Problem Solving

Understand

1. What is the problem about?

2. What information is given in the problem?

3. What is the question?

Plan

4. What problem-solving strategies might I try to help me solve the problem?

5. About what do I think my answer will be?

Solve

6. How can I solve this problem?

7. How can I state my answer in a complete sentence?

Check

8. How do I know whether my answer is reasonable?

9. How else might I have solved this problem?

Problem Solving Skill 1: Choose the Operation

The American flag has 13 stripes. There are 7 red stripes. The rest are white. How many white stripes are there?

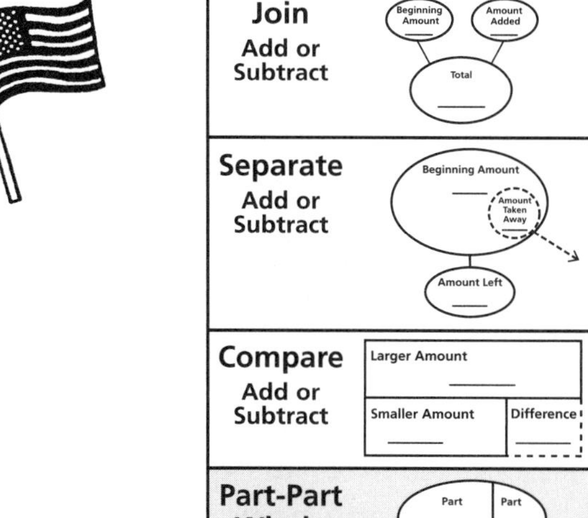

UNDERSTAND

Reread the problem.
- Underline what you need to find.
- Circle the facts.

PLAN

Think about what you know.
- You know the whole, 13 stripes.
- You know one part, 7 red stripes.

Think about what operation to use.
To find the missing part, subtract.

SOLVE

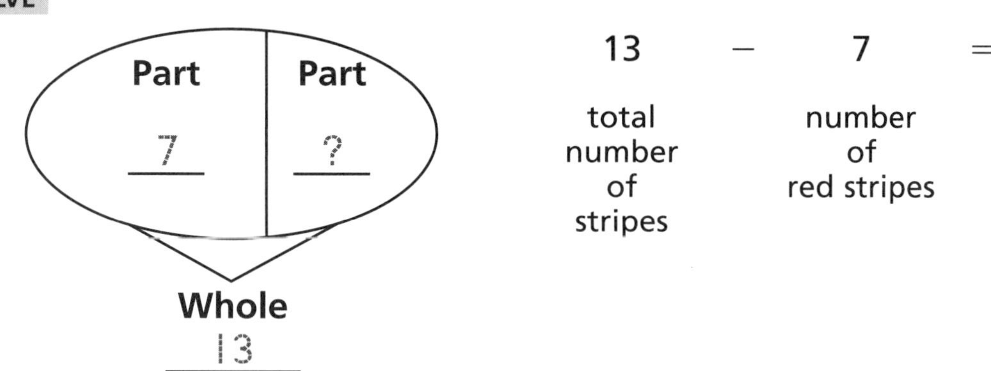

Part	Part
7	?

Whole
13

$$13 - 7 = 6$$

total number of stripes	number of red stripes	number of white stripes

So, there are ___6___ white stripes.

CHECK

- How can you use addition to check your answer?

▶ Try These

Write whether you would *add* or *subtract*. Then solve.

1. One day Leah hiked 12 miles. Another day she hiked 13 miles. How many miles did she hike in all?

2. Leah took 15 pictures on her trip. She lost 5 pictures. How many pictures does she have left?

PRACTICE ON YOUR OWN

Write whether you would *add* or *subtract*. Then solve.

1. Karen baked 48 loaves of bread. She sold 30 loaves. How many loaves of bread does she have left?

2. Kay planted 25 tulips and 30 roses. How many more roses did she plant than tulips?

3. Michael is having a party. He made 12 prize bags for his friends. He has 8 more to make. How many bags will he make in all?

4. Giraffes can run 40 miles per hour. Cheetahs can run 70 miles per hour. How much faster can cheetahs run than giraffes?

5. Samantha has 54 baseball cards. She gets 10 more from her brother. How many baseball cards does she have in all?

6. There are 25 students in Mr. Joy's class. There are 20 students in Mrs. White's class. How many students are there in both classes?

7. Jon took a trip. In June he visited 10 states. In July he visited 13 states. How many states did he visit in all?

8. Tony has 35 seashells. 15 seashells are red and the rest are blue. How many blue seashells does he have?

▶ Quiz

Write whether you would *add* or *subtract*. Then solve.

9. A large pizza has 16 slices. 10 slices are cheese and the rest are ham. How many slices are ham?

10. Fran had 37 marbles. She bought 12. How many marbles does she have now?

11. Joe and Tim wrote a book together. Joe wrote 45 pages. Tim wrote 30 pages. How many pages did they write in all?

12. On Monday, Mary rode her bike 32 miles. On Tuesday, she rode 20 miles. How many more miles did she ride on Monday?

Problem Solving Strategy 2:
Use Logical Reasoning

Dan used 1 hundred block and 5 tens to model 150. What is another way he can model 150?

UNDERSTAND

Reread the problem.
- Underline what you need to find.
- Circle the facts.

PLAN

- What strategy can you use to solve the problem?
- Which block will Dan have to trade?

SOLVE

How can you use the strategy to solve the problem?

Step 1
Start with Dan's blocks.

Step 2
Trade the hundred block for 10 tens.

Step 3
Count the number of tens Dan used.

So, another way to model 150 is __15__ tens.

CHECK

- Explain how to model 150 a third way.

▶ Try These

Use logical reasoning and solve.

1 Sam modeled a number using 10 tens and 5 ones. What number did he model?

2 Tara modeled a number with 1 hundred and 10 tens. What number did she model?

Name ____

PRACTICE ON YOUR OWN

Use logical reasoning and solve.

1. How can you model 139 using only hundreds and ones?

2. Harry wants to model 415. How can he do this?

3. Vicky has 32 ones. What is another way she could model the same number?

4. How can Colin model 125 without using tens blocks?

5. Joe has 2 hundreds and 10 tens. He wants to model 250. How can he do this?

6. Paul modeled a number using 2 hundreds, 1 ten, and 3 ones. What is the number?

7. Anne has 1 hundred block, 12 tens, and 5 ones. How can she model the number 221?

8. Emma used 34 tens and 8 ones to model a number. What number did she model?

9. What is another way to model 10 tens?

10. Dolly used 14 tens. What number did she model?

11. Ben has 5 tens and 2 ones. Kim has 6 tens. What number can they model if they put all their blocks together?

12. How can you model the number 148 using only tens and ones?

Quiz

Use logical reasoning and solve.

13. José has 23 ones. Cindy has 9 tens. What number can they model if they put their blocks together?

14. How can Fred model 130 without using tens blocks?

15. What is another way to model 20 tens?

16. Julie used 9 tens and 6 ones. What number did she model?

Problem Solving Skill 3: Use a Bar Graph

Wendy wants to climb a mountain higher than 3,100 meters. Which Sierra Nevada mountains are higher than 3,100 meters?

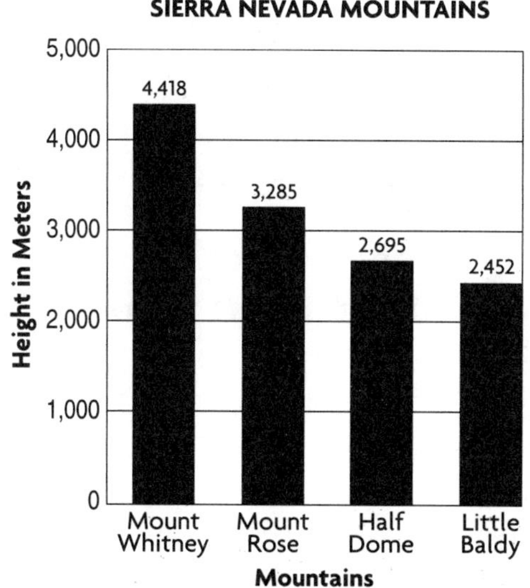

UNDERSTAND

- Underline what you need to find.
- Circle the facts.

PLAN

Compare the heights of the bars and find out how high each mountain is.

SOLVE

Step 1 How are the mountains organized in the bar graph?

Step 2 List each mountain and its height.

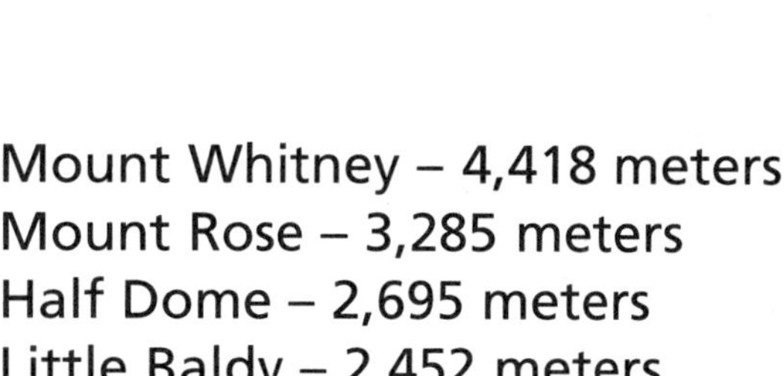

Mount Whitney – 4,418 meters
Mount Rose – 3,285 meters
Half Dome – 2,695 meters
Little Baldy – 2,452 meters

Step 3 Find the mountains that are higher than 3,100 meters.
- Compare the digits in the thousands place. Find those with a 3 or greater.
- Compare 3,285 to 3,100.   3,285 > 3,100
- Compare 4,418 to 3,100.   4,418 > 3,100

So, _____Mount Rose_____ and _____Mount Whitney_____ are higher than 3,100 meters.

CHECK

- How do you know that your answer is reasonable?

▶ Try These

Use the bar graph above.

1 About how much higher is Mount Whitney than Little Baldy?

2 Which mountain is lower than Half Dome?

Name ____

PRACTICE ON YOUR OWN

Use the bar graph at the right.
Victor is studying the Great Lakes.
He made this bar graph to show
how deep each lake is.

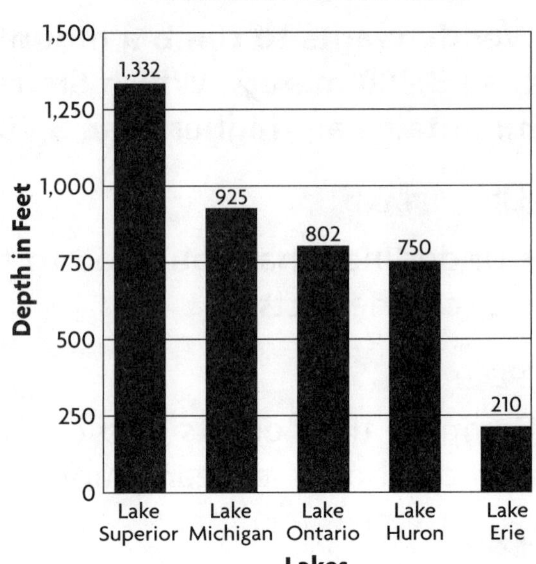

1 Order the depths of the lakes from deepest to most shallow.

2 Which lakes are deeper than Lake Ontario?

3 Which is most shallow?

4 Which lake is the deepest?

5 What is the difference in depth of the deepest and most shallow lakes?

6 Which lake is deeper than Lake Ontario but not as deep as Lake Superior?

▶ Quiz

Use the bar graph at the right.

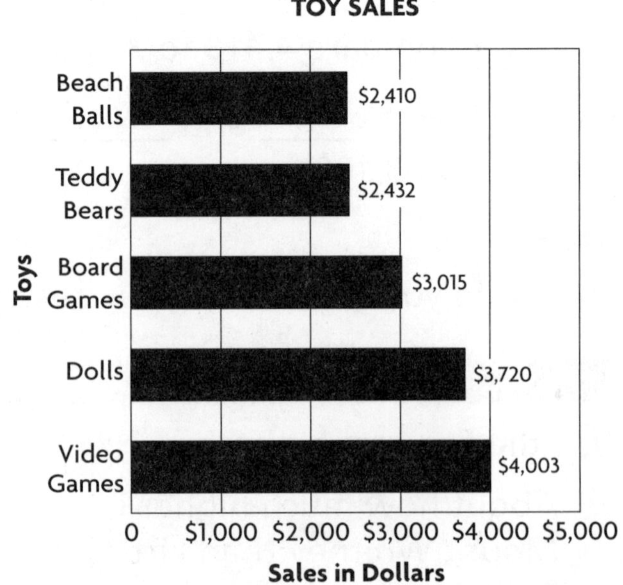

7 Which toys had sales of about the same amount?

8 Which toy sold the most?

9 Which toy had more sales than board games and fewer sales than video games?

10 Which toys had more than $3,000 in sales?

IPS6 Intervention • Problem Solving

Problem Solving Strategy 4:
Predict and Test

The animal shelter has 13 cats and dogs. There are 5 more cats than dogs. How many dogs are at the animal shelter?

UNDERSTAND

Reread the problem.

- Underline what you need to find.
- Circle the facts.

PLAN

You know how many animals there are in all. You do NOT know how many are dogs and how many are cats. You can use *predict and test* to find out.

SOLVE

You know that there are 5 more cats than there are dogs. Start with a guess about dogs because you know you have to add 5 to that number.

DOGS	CATS	DOGS + CATS	NOTES
5	$5 + 5 = ⑩$	$5 + 10 = 15$	Too high
3	$3 + 5 = ⑧$	$3 + 8 = 11$	Too low
4	$4 + 5 = ⑨$	$4 + 9 = 13$	Just right!

So, there are __4__ dogs at the animal shelter.

CHECK

- How did the first two predictions lead to the right answer?

▶ Try These

Use *predict and test* to solve.

1 What if there were only 3 more cats than dogs. What is the new number of cats?

2 Amy is 4 years older than Andy. The sum of their ages is 14. How old is Andy?

Name ____

PRACTICE ON YOUR OWN

Use *predict and test* to solve.

1 The sum of two numbers is 10. Their difference is 4. What are the two numbers?

2 The sum of two numbers is 12. Their difference is 2. What are the two numbers?

3 Mario has 45 yellow and red balls. There are 5 more red balls than yellow. How many balls does he have of each color?

4 Maria and Roy have 10 pets altogether. Roy has 2 fewer pets than Maria. How many pets does Maria have?

5 Jacob and Kara ran a total of 20 laps in a race. Kara ran 4 more laps than Jacob. How many laps did Jacob run?

6 Eva baked 35 cookies. She made 15 more sugar cookies than chocolate. How many sugar cookies did she make?

7 Rich bought 18 pieces of fruit. He bought 2 more oranges than peaches. How many peaches did he buy?

8 Jill and Kim saved $25. Jill saved $7 more than Kim. How many dollars did Kim save?

▶ Quiz

Use *predict and test* to solve.

9 The Blakes have 14 chicks and rabbits. There are 6 more chicks than rabbits. How many chicks are there?

10 The sum of two numbers is 15. Their difference is 1. What are the two numbers?

11 Marie bought 19 cans of food. She bought 5 more cans of peas than corn. How many cans of peas did she buy?

12 The sum of Brad and Julio's ages is 24. Brad is 2 years younger than Julio. How old is Julio?

Problem Solving Skill 5:
Estimate or Exact Answer

Daniel climbed 510 meters up Mount Sunflower on Monday. He climbed 680 meters on Tuesday. About how many meters did he climb in all?

UNDERSTAND

Reread the problem.
- Underline what you are to find.
- Circle the facts.

PLAN

The question asks *about* how many meters.
So, all you need is an estimate for your answer.

SOLVE

Estimate the sum by rounding each addend to the nearest hundred. Use the join Math Map.

    510 ➜ 500
    680 ➜ 700

   500 + 700 = 1,200

So, Daniel climbed *about* __1,200__ meters.

CHECK

- How do you know that your estimate is reasonable?

▶ **Try These**

Tell whether you need an exact answer or an estimate. Then solve.

1. There are 16 boys and 18 girls in Mr. Phillip's class. How many students are in his class?

2. If one towel costs $5.99, about how much would 3 towels cost?

Name _______________________________

PRACTICE ON YOUR OWN

**Tell whether you need an exact answer or an estimate.
Then solve. For 1–2, use the table.**

BOOK SALES IN NOVEMBER	
Week 1	$793.00
Week 2	$557.00
Week 3	$689.00
Week 4	$803.00

1 About how much money was earned in all during Week 1 and Week 2?

2 How much more money was earned in Week 4 than in Week 2?

3 Kim has $10.00. She wants to buy a toy that costs $7.85 and a book that costs $3.25. Does she have enough money?

4 There are 112 students at the park. About 40 of them are in third grade. How many students are not in third grade?

5 Mr. Owen sold 18 kites. He has 60 kites to sell in all. About how many more kites does he need to sell?

6 Sue bought 2 paintings. One cost $45. The other cost $98. About how much did both paintings cost?

7 Tim had a yard sale. One day he made $27.50. The next day he made $32.75. How much money did he make in all?

8 Lenny can play 45 minutes before dinner. He has been playing for 15 minutes. How much time does he have left to play?

▶ Quiz

**Tell whether you need an exact answer or an estimate.
Then solve.**

9 Peter has 66 beans. He gives 32 to his friends. About how many beans does he have left?

10 Dan has $145. He wants to buy a bike that costs $355. How much more money does he need?

11 Gina scored 21 points in one basketball game and 17 points in another game. How many points did she score in all?

12 A movie is 181 minutes long. Sandy has been watching it for 75 minutes. About how much time is left in the movie?

IPS10 Intervention • Problem Solving

Problem Solving Strategy 6: Make a Table

Erin has two $1 bills, 3 quarters, 4 dimes, and 3 nickels. How many different ways can she pay for a card that costs $1.50?

UNDERSTAND

Reread the problem.
- Underline what you are to find.
- Circle the facts.

PLAN

What strategy can you use to solve the problem?

SOLVE

How can you use the strategy to solve the problem?

$1 BILLS	QUARTERS	DIMES	NICKELS	TOTALS
1	2			$1.50
1		4	2	$1.50
1	1	1	3	$1.50
1	1	2	1	$1.50

So, there are ___4___ different ways for Erin to make $1.50.

CHECK

- Is your answer reasonable? Explain.

▶ Try These

Make a table to solve.

1. Erin wants to buy a book that costs $2.40. How many different groups of bills and coins can she use?

2. If Erin uses only bills and dimes to pay for the book, how many dimes will she use?

PRACTICE ON YOUR OWN

Make a table to solve.

1 Eddie has one $1 bill, 2 quarters, 2 dimes, and 5 nickels. How many ways can he pay for a ball that costs $1.70?

2 If Eddie uses the most coins possible to pay for the ball, how many quarters will he use?

3 Kendra has 3 quarters, 5 dimes, and 5 pennies. How many ways can she pay for a pen that costs $0.75?

4 Kendra needs to save one quarter for the bus. Which way will she not pay for the pen?

For 5–8, use the table.

5 Jimmy has 1 quarter, 5 dimes, and 5 pennies. How many ways can he pay for a hook?

6 Anna has only quarters and dimes. If she uses 6 coins to pay for the line, what coins does she use?

FISHING EQUIPMENT	
Pole	$5.25
Hook	$0.25
Bait	$2.75
Line	$1.50

7 Mario has three $1 bills, 4 quarters, 1 dime, and 6 nickels. How many ways can he pay for the bait?

8 Is it possible for Mario to pay for the bait using no quarters?

▶ Quiz

Make a table to solve.

9 A hot dog costs $1.65. How many ways can you pay for the hot dog with 5 quarters, 6 dimes, and 3 nickels?

10 How can you pay for the hot dog with the fewest coins?

11 Li has 3 quarters, 3 nickels, and 10 pennies. How many ways can she buy a cookie for $0.85?

12 Can Li pay for the cookie with only her nickels and pennies? Explain.

Problem Solving Skill 7:
Sequence Events

Arthur wanted to know if Nintendo was introduced before or after Connect Four.

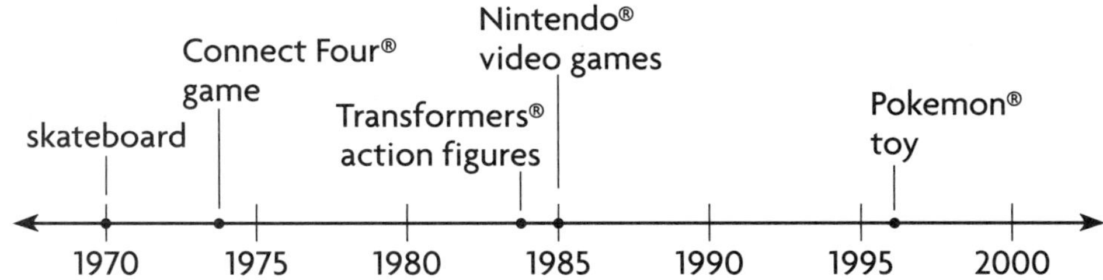

UNDERSTAND

Reread the problem.
- Underline what you are to find.
- Circle the facts on the time line.

PLAN

A **time line** is a drawing that shows when and in what order events took place. It is read from left to right. The dates at the left happened first. The dates at the right happened later.

SOLVE

- Nintendo is to the right of Connect Four.
- Compare the dates. 1974 < 1985

So, Nintendo was introduced _____after_____ Connect Four.

CHECK

- Suppose you had no time line, but you knew when the toys were introduced. How could you tell which toy was introduced first?

▶ Try These

Use the time line to solve.

1. Boogie® Boards were introduced in 1980. Explain whether the point for Boogie Boards would go to the left or right of Pokemon.

2. Were Transformers or the skateboard introduced first?

PRACTICE ON YOUR OWN

Use the time line to solve.

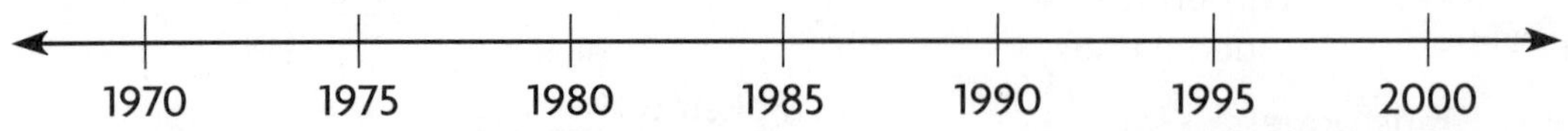

1. Add a point to the time line for each toy listed in the table. Label each point with the name of the toy.

INTRODUCTION OF TOYS	
Rubik's Cube® puzzle	1974
Game Boy® game	1989
Furby® animal	1998
Stretch Armstrong® figure	1976
Koosh® ball	1987

2. Hello Kitty® was introduced in 1975. Between which two toys should it be placed?

3. Which toy was introduced before Furby but after the Koosh Ball?

4. List these toys in order from newest to oldest: Koosh Ball, Furby, Game Boy.

5. Which toy was introduced after the Game Boy?

6. Which toy was introduced first: Furby or the Rubik's Cube?

▶ Quiz

Use the time line to solve.

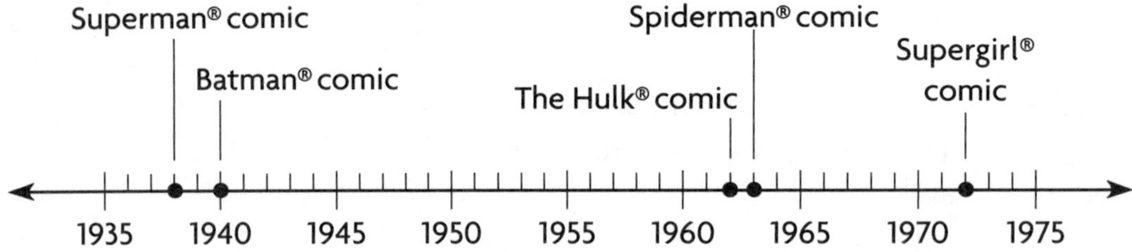

7. List these comics in order from earliest to latest: *Supergirl, The Hulk, Batman*.

8. *Wonder Woman®* was made in 1941. Between which two comics should it be placed?

9. Which comics were made before *Spiderman* but after *Superman*?

10. Which comic was made after *Spiderman*?

Problem Solving Skill 8:
Too Much/Too Little Information

Two students went to the art store. A can of paint costs $3.00. Each student bought 4 cans of paint. How many cans of paint did the students buy in all?

UNDERSTAND

Reread the problem.
- Underline what you are to find.
- Circle the facts.

PLAN

- What facts do you need to solve this problem?

- Is there information in this problem that you do not need? If so, what is it?

SOLVE

_2_ students × _4_ cans each = _8_ cans

_2_ × _4_ = _8_

So, there are _8_ paint cans in all.

CHECK

- Suppose the students also bought 4 paintbrushes. What information do you need to find how much they spent on paint and paintbrushes?

▶ Try These

Write *too much, too little,* or the *right amount* of information for each problem. Solve those with too much or the right amount of information. Tell what is missing for those with too little information.

1. You have $10 to spend on a ruler and tape. The tape costs $4. Do you have enough to buy both?

2. Ricky bought 2 boxes of pens. Each box costs $3. Each box has 8 pens. How much did Ricky pay for 2 boxes of pens?

Name ____________________________________

PRACTICE ON YOUR OWN

Write *too much, too little,* or the *right amount* of information
for each problem. Solve those with too much or the right
amount of information. Tell what is missing for those
with too little information.

1. Brandon had $8. He bought stickers for $2 and scissors for $3. How much did he spend?

2. An eraser costs $1. Glue costs $5. How much more does the glue cost than the eraser?

3. Teresa bought 2 books. They cost $6 and $7. How much change will she receive from a $20 bill?

4. A box of pencils holds 10 pencils. Each box costs $2. How much do 6 boxes cost?

5. Roger walked 7 blocks to the bookstore. He walked 5 more blocks to the art store. This took him 30 minutes. How many blocks did he walk in all?

6. Ashley has $15 for the book fair. She wants to buy 5 books. Does she have enough money for all 5?

▶ **Quiz**

Write *too much, too little,* or the *right amount* of information
for each problem. Solve those with too much or the right
amount of information. Tell what is missing for those
with too little information.

7. Julie bought some pens. Each pen costs $1. How much money did she spend on pens?

8. Kevin bought a school bag for $10 and a lunch box for $7. He paid with a $20 bill. How much did he spend in all?

9. Angie has 18 pencils. She gives 6 to her brother. How many pencils does she have left?

10. Val wants to buy 3 books. She has $9. Does she have enough to buy them all?

Problem Solving Strategy 9: Find a Pattern

Kara wrote this number pattern: 1, 2, 4, 5, 7, 8, 10, 11, and 13. What is a rule for her pattern? What are the next four numbers she will write?

UNDERSTAND

Reread the problem.
- Underline what you are to find.
- What information will you use?

PLAN

You can use the strategy *find a pattern* to solve the problem.

SOLVE

- Use a number line to find the pattern.
- Write a rule for the pattern. Follow a rule to write the next four numbers.

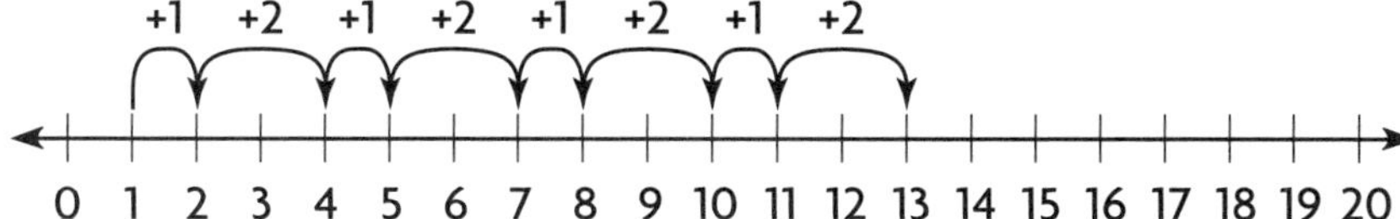

A rule is to add __1__ and then add __2__.
The next four numbers are __14__, __16__, __17__, and __19__.

CHECK

- How can you find the rule without using a number line?

 **Try These**

Use *find a pattern* to solve.

1 Martin made this number pattern: 16, 14, 12, and 10. What is the rule for his pattern?

2 What are the next four numbers in Martin's pattern?

PRACTICE ON YOUR OWN

Use *find a pattern* to solve.

1. Tor thought of this pattern: 2, 4, 6, 8 and 10. What is a rule for his pattern?

2. Could the number 25 be in Tor's pattern? Why?

3. Joe made this pattern: 1, 5, 9, 13, 17, and 21. What is a rule for his pattern? What are the next four numbers?

4. Stan's pattern is 5, 10, 15, and 20. What is a rule for his pattern? What are the next four numbers?

5. Dolly made a pattern that followed this rule: multiply by 2. She started with 1. What are the next four numbers in her pattern?

6. Miss Sanchez wrote this pattern on the board: 4, 3, 5, 4, 6, 5, 7, 6, 8, 7, and 9. What is a rule for her pattern?

7. Write 5 numbers in a pattern that follows Miss Sanchez's rule. Start with 2.

8. Ting's rule is add 3, subtract 1. Write five numbers in her pattern. Start with 2.

▶ Quiz

Use *find a pattern* to solve.

9. Brenda made a pattern. The first four numbers are 3, 5, 7, and 9. What is a rule for her pattern? What are the next four numbers?

10. Jake made a pattern that started with 3. The rule he used was add 4, subtract 1. What are the next four numbers in his pattern?

11. Phil's pattern is 10, 20, 30, 40 and 50. What is a rule for his pattern? What are the next four numbers?

12. Helen wrote this pattern: 24, 21, 18, and 15. What is a rule for her pattern? What are the next four numbers?

Problem Solving Skill 10:
Use a Pictograph

Jan read in her science book that some planets have more
than one moon. How many moons does Mars have?

NUMBER OF MOONS AROUND PLANETS	
Neptune	○ ○ ○ ○
Mars	○
Earth	◖
Key: Each ○ = 2 moons.	

UNDERSTAND

- Underline what you are to find.
- Circle the facts.

PLAN

Find Mars in the pictograph and use the key to count its moons.

SOLVE

Look at the key. Each ○ equals 2 moons.
Use multiplication to find the number of moons for Mars.

1	×	2	=	2 moons
Number of circles		Each circle stands for 2 moons.		Total number of moons for Mars

So, Mars has __2__ moons.

CHECK

- What if the key told you that each circle was equal to 1 moon?
  How could you find the number of moons on Mars?

▶ Try These

For 1–2, use the pictograph.

1. Which planet in the pictograph
   has the most moons? How
   many moons does it have?

2. How many more moons does
   Neptune have than Mars?

PRACTICE ON YOUR OWN

For 1–4, use the pictograph.
Constellations are pictures in the sky made out of stars.

1 How many stars does the ⭐ next to Pisces represent?

2 Which constellation has the fewest stars? How many stars does it have?

3 How many stars make up Taurus?

4 How many stars do Aries and Leo have together?

CONSTELLATIONS	
Name	**Number of Stars**
Leo the Lion	★ ★ ★ ★ ★ ⯨
Aries the Ram	★ ★
Taurus the Bull	★ ★ ★ ★ ★ ★ ★ ⯨
Pisces the Fish	★ ⯨
Key: Each ★ = 2 stars.	

▶ **Quiz** ....

For 5–10, use the pictograph.

5 How many students are in the 3rd grade?

6 Which animal was reported on by 12 third graders?

7 Which animal was reported on the most? How many students reported on that animal?

8 How many more students reported on alligators than bears?

3rd GRADE ANIMAL REPORTS	
Kangaroos	👤 👤 👤
Alligators	👤 👤 👤 👤
Bears	👤 👤
Rabbits	👤
Key: Each 👤 = 4 students.	

IPS20 Intervention • Problem Solving

Problem Solving Skill 11: Multistep Problems

In football a touchdown is 6 points and a field goal is 3 points. A team made 2 touchdowns and 2 field goals. How many points did they score?

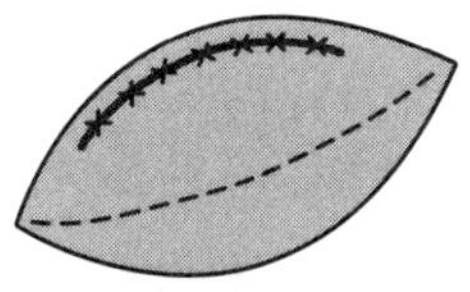

UNDERSTAND

Reread the problem.
- Underline what you are to find.
- Circle the facts.

PLAN

Multiply to find how many points the team scored in touchdowns and how many points they scored in field goals. Then *add* to find how many points they scored in all.

SOLVE

Step 1	**Step 2**	**Step 3**
How many points did the team score in touchdowns?	How many points did the team score in field goals?	How many points did the team score in all?
$2 \times 6 = 12$	$2 \times 3 = 6$	$12 + 6 = 18$

So, the team scored __18__ points.

CHECK

- If the team scored 2 touchdowns and 3 field goals, how many points would they score in all?

▶ Try These

Solve.

1. Li, Jill, and Pat have 31 cards in all. Li has 18 cards. Pat has 5 cards. How many cards does Jill have?

2. Jo buys 4 pens and Ian buys 3 pens. Each pen costs $2. How much do they pay in all?

PRACTICE ON YOUR OWN

Solve.

1. Tom had 4 boxes of books. Each box held 5 books. He gave 10 books to his sister. How many books does he have left?

2. Martha bought 3 bunches of purple flowers and 4 bunches of pink flowers. Each bunch costs $5. How much did she spend in all?

3. Amir bought 5 bags of blueberry muffins and 3 bags of bran muffins. Each bag held 6 muffins. How many muffins did he buy in all?

4. A can of red paint costs $5. A can of gold paint costs $6. Jan bought 3 cans of red paint and 2 cans of gold paint. How much did she spend in all?

5. Annie is reading a book that is 150 pages long. One night she read 45 pages. The next night she read 65 pages. How many pages does she have left to read?

6. Jessie had a box of 65 markers. She gave 12 to her friend and 8 to her brother. How many markers does she have left?

▶ Quiz

Solve.

7. Mark bought 2 small pizzas and 2 medium pizzas. A small pizza has 8 slices and a medium pizza has 10 slices. How many slices of pizza did he buy in all?

8. Kim bought movie tickets for her whole family. She bought 3 children's tickets for $5 each. She bought 2 adult tickets for $7 each. How much did she spend in all?

9. Pilar wants to make 175 bracelets to sell. One week she made 25. Another week she made 20. How many bracelets does she have left to make?

10. Wilson has 4 bags of marbles. Each bag holds 8 marbles. If he gives 12 marbles to his brother, how many marbles will he have left?

Problem Solving Strategy 12:
Write a Number Sentence

Molly puts 20 trading cards in her binder. She puts 5 cards on each page. How many pages will Molly need for her trading cards?

UNDERSTAND

Reread the problem.

- Underline what you are to find.
- Circle the facts.

PLAN

You can *write a number sentence* to find the number of pages Molly will need. You can use the Math Map for equal groups.

SOLVE

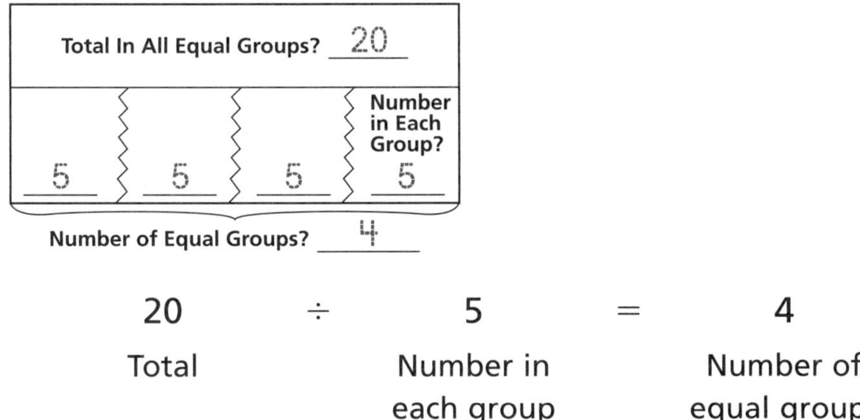

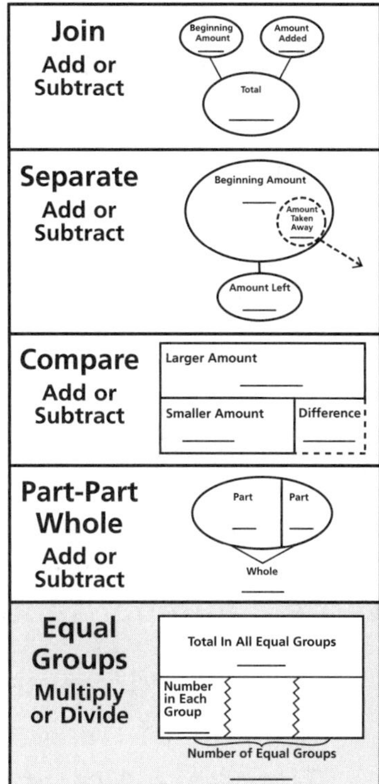

20	÷	5	=	4
Total		Number in each group		Number of equal groups

So, Molly needs __4__ pages for her cards.

CHECK

- How can you use multiplication to check your answer?

▶ Try These

Write a number sentence to solve.

1 If Molly buys 10 more cards to put in her binder, how many pages will she need for all her cards?

2 Lucia has 16 cards to divide equally among her 4 friends. How many cards will each friend get?

PRACTICE ON YOUR OWN

Write a number sentence to solve.

1 Omar has 3 bags of marbles. There are 7 marbles in each bag. How many marbles does he have in all?

2 Shiva divided 18 shells equally between herself and her sister. How many shells did each girl get?

3 Sean wants to put 36 cards in his binder. Each page holds 6 cards. How many pages will Sean use?

4 Tara's binder has 10 pages. Each page holds 6 cards. How many cards can Tara's binder hold in all?

5 Eli made 8 trays of ice pops. He made 48 ice pops in all. How many ice pops were in each tray?

6 Ian hiked 80 miles in 8 days. He hiked the same distance each day. How many miles did he hike each day?

▶ **Quiz** ....

Write a number sentence to solve.

7 Zoë scored 7 points in each quarter of her basketball game. There are 4 quarters in a basketball game. How many points did Zoë score?

8 There are 5 tables of students in Mrs. March's class. There are 5 students at each table. How many students are in the class?

9 Trey made 64 cookies. 8 cookies fit on one cookie sheet. How many cookie sheets did he use?

10 Anita made 5 bracelets for her friends. Each bracelet had 9 beads. How many beads did Anita use?

11 Jason picked 35 oranges from 5 trees. He picked the same amount from each tree. How many oranges did he pick from each tree?

12 Sonya planted 16 tulip bulbs in 4 rows in her yard. How many bulbs did she plant in each row?

Problem Solving Skill 13: Choose the Operation

On a hike, Gil saw 3 bird's nests. There were 5 birds in each nest. How many birds did he see in all?

UNDERSTAND

Reread the problem.

- Underline what you are to find.
- Circle the facts.

PLAN

You know that there are 3 nests with 5 birds in each. You need to find how many birds in all. You are joining 3 equal groups, so you multiply.

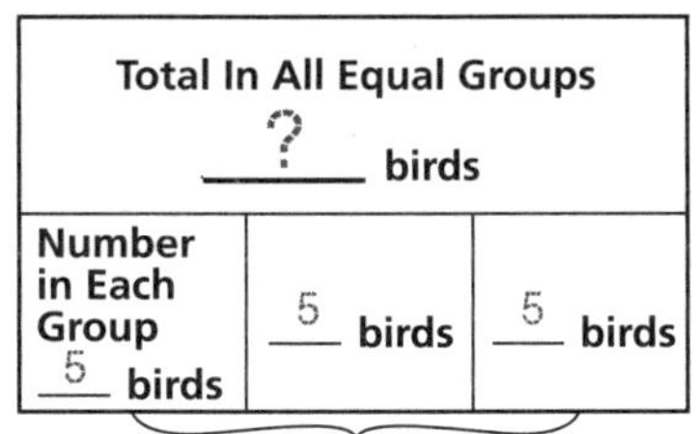

SOLVE

Write an equation:

3	×	5	=	15
Number of groups		Number in each group		Total in all equal groups

So, Gil saw __15__ birds in all.

CHECK

- How can you use addition to check your answer?

▶ Try These

Choose the operation. Write an equation. Then solve.

1 12 campers slept in tents. There were 4 campers in each tent. How many tents were there?

2 June counted 10 deer. 5 walked away. How many deer are left?

PRACTICE ON YOUR OWN

Choose the operation. Write an equation. Then solve.

1. Ella made sandwiches. 3 were tuna, 4 were cheese and 6 were ham. How many sandwiches did she make in all?

2. Juana saw 4 families of ducks. She counted 5 ducks in each family. How many ducks did she see in all?

3. Cameron cooked 2 hot dogs for each of his 7 friends. How many hot dogs did he cook in all?

4. Olivia divided 16 stones into 4 equal groups. How many stones were in each group?

5. While hiking, Erik saw 5 deer, Sara saw 8, and Mitch saw 6. How many deer did they see in all?

6. Lily counted 5 frogs and 17 fish. How many more fish did she see than frogs?

7. Hailey uses 20 lemons to make lemonade. She uses 2 lemons for each glass. How many glasses does she make?

8. A large popcorn costs $5.75, and a small popcorn costs $1.25. How much more does the large popcorn cost than the small one?

▶ Quiz

Choose the operation. Write an equation. Then solve.

9. There are 3 groups of hikers. Each group has 3 hikers. How many hikers are there in all?

10. Russ caught 19 crickets. 5 of them ran away. How many crickets are left?

11. Zack saw 5 shooting stars. Rick saw 3, and Liz saw 6. How many shooting stars did they see in all?

12. William has 8 crackers. He divides them into groups of 2. How many groups of crackers did he make?

Problem Solving Strategy 14: Work Backward

Tom baked 2 batches of cookies. The extra dough made 3 more cookies. He made 15 cookies in all. How many cookies were in each batch?

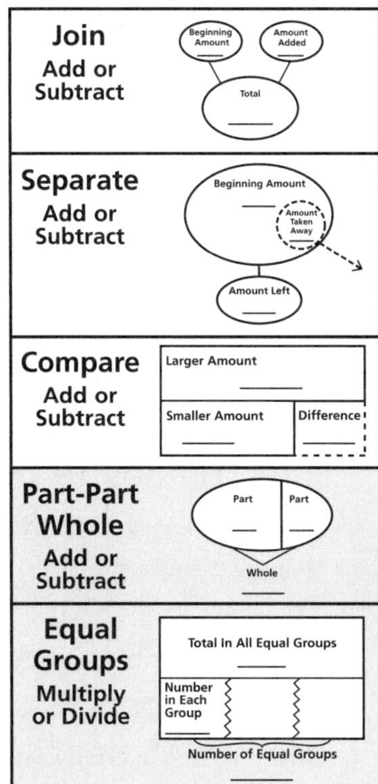

UNDERSTAND

- Underline what you are to find.
- Circle the facts.

PLAN

Think about the problem and decide which strategy to use. You can use the facts in the problem. *Work backward* through the facts to find the number of cookies in each batch.

SOLVE

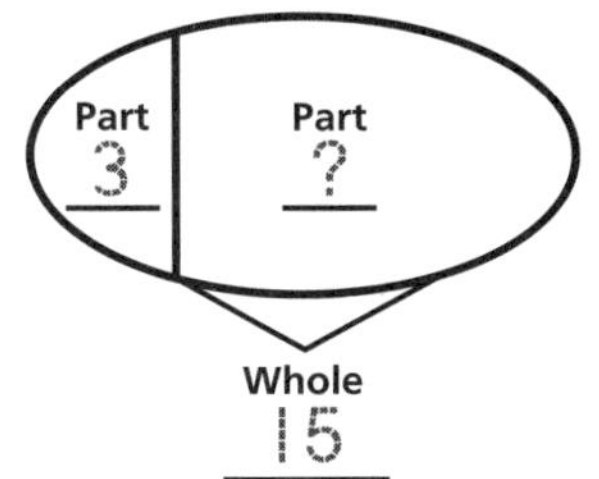

Total In All Equal Groups
12

Number in Each Group
?   ?

Number of Equal Groups
2

15 cookies in all − 3 extra cookies = 12 cookies in 2 batches.
15 − 3 = 12
There are 12 cookies in 2 batches.

Divide to find the number of cookies in each batch.
12 ÷ 2 = 6
So, there are _6_ cookies in each batch.

CHECK

- How can you check your answer?

▶ Try These

Work backward to solve.

1 Al baked 3 batches of rolls and made 4 extra rolls. He made 28 rolls in all. How many rolls were in each batch?

2 Li made 4 batches of rolls. She made 1 extra roll. She made 41 rolls in all. How many rolls were in each batch?

Name ____

PRACTICE ON YOUR OWN

Work backward to solve.

1. Jan spent $25 at the music store. She bought a walkman for $10 and 3 CDs. How much did each CD cost?

2. Ian spent $7 at the store. He bought 5 cookies. He bought popcorn for $2. How much did each cookie cost?

3. Sam has 4 boxes of books. He has 3 books that are not in a box. He has 19 books in all. How many books are in each box?

4. Mrs. Butler made 3 boxes of rolls. She had 4 extra rolls. She made 34 rolls in all. How many rolls were in each box?

5. Jake had 5 boxes of pencils. He gave 10 pencils to his brother. He has 15 pencils left. How many pencils are in each box?

6. Celia had 8 packs of juice. She gave 4 cans to her friend. She has 20 cans left. How many cans of juice are in one pack?

7. Dylan bought 2 board games that cost the same amount. He gave the cashier $40 and received $4 in change. How much money did each game cost?

▶ Quiz

Work backward to solve.

8. Jim had 6 cans of tennis balls. He lost 6 balls. He has 18 balls left. How many balls were in each can?

9. Bi spent $15 at the movie theater. She bought popcorn for $3 and 2 tickets. How much did each ticket cost?

10. Aisha spent $30 at the pet store. She bought cat food for $10 and 2 bags of dog food. How much did each bag of dog food cost?

11. Frank bought 2 hats that cost the same price. He paid with a $20 bill. He got $8 in change. How much did each hat cost?

Problem Solving Strategy 15: Make a Table

Dave and June used the spinners at the right. They spun the pointers 10 times. Each time, they wrote down the sum of the two numbers they spun. Their results were 4, 4, 2, 5, 4, 6, 4, 2, 3, and 4. Which sum occurred most often?

$$2 + 2 = 4$$

UNDERSTAND

Reread the problem.
- Underline what you are to find.
- Circle the facts.

PLAN

Think about the problem and decide which strategy to use.
Since you know the sums, you can *make a table* to organize the data.

SOLVE

How can you make a table?

Step 1: Put the sums in one column.

Step 2: In the other column, make a tally mark for each time the sums were spun.

So, the sum 4 occurred most often.

SPINNER EXPERIMENT	
Sums	**Number of Spins**
2	I I
3	I
4	卌
5	I
6	I

CHECK

- Is your answer reasonable?

▶ Try These

Solve.

1 Dave and June spun the pointers 3 more times. They spun 2, 3, and 4. How many 2's have they spun in all?

2 If Dave and June spin the pointers 15 times and record the sums in a table, how many tallies should there be?

PRACTICE ON YOUR OWN

Solve.

For 1–3, use the spinners.

1 Ting and Bo spun the pointers of these spinners and recorded the sums of their spins. They were 8, 8, 9, 7, 6, 9, 8, 7, 8, and 10. Make a table to show the sums.

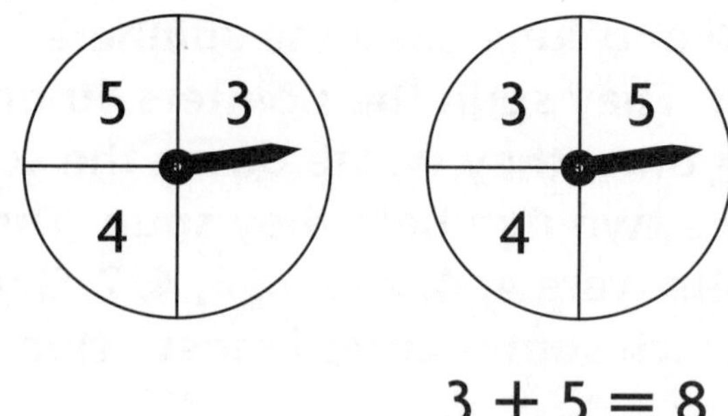

$$3 + 5 = 8$$

2 Which sum occurred most often?

3 How many tally marks are in your table?

4 Roll two number cubes 15 times to find out what sum you will roll most often. Make a table to show your sums.

5 When you roll two number cubes, what is the greatest possible sum?

6 How many tally marks are in your table?

7 Which sum(s) did you roll least often?

8 If you roll three number cubes, what is the greatest possible sum? the least possible sum?

▶ Quiz

Solve.

9 Roll two number cubes 10 times. Subtract the lesser number from the greater number. Make a table to show your differences.

10 Which difference(s) did you roll least often? most often?

11 How many tally marks should be in your table for this experiment?

12 You roll the two number cubes 8 more times. How many tally marks are in your table now?

IPS30 Intervention • Problem Solving

Problem Solving Strategy 16:
Make a Graph

**The soccer team sold candy bars. Gina sold
10 bars, Sam sold 8, Nat sold 20, and Kelly sold 4.
What is one way to show the candy bar sales in
a graph?**

UNDERSTAND

Reread the problem.
- Underline what you are to find.
- Circle the facts.

PLAN

You know how many candy bars each person sold.
So, you can *make a pictograph* to show these numbers.

SOLVE

Step 1: Give your graph a **title**.

Step 2: Write a name in each
row as a **label**.

Step 3: Choose a **key**. The key
tells how many candy bars each
picture stands for.

Step 4: Decide how many
pictures should be next to
each name.

CANDY BARS SOLD	
Gina	🍫🍫🍫🍫🍫
Sam	🍫🍫🍫🍫
Nat	🍫🍫🍫🍫🍫🍫🍫🍫🍫🍫
Kelly	🍫🍫
Key:	🍫 = 2 candy bars

CHECK

- How can you check your answer?

▶ **Try These**

Make a graph for each problem.

1 Tim, May, and Liz also play
soccer. Tim sold 6 candy bars,
May sold 12, and Liz sold 15.
Make a pictograph to show
the data.

2 Use a different key to draw
a new pictograph for the
information in Problem 1.

PRACTICE ON YOUR OWN

Make a graph for each problem.

1 Andy, Will, and Art read books. Andy read 20 books, Will read 15, and Art read 10. Use a key of 5. Make a pictograph to show the data.

2 Dolly read 35 books. Using the key from Problem 1, how many pictures would you draw to show how many books she read?

For 3–8, use the data in the table.

3 Julie made a pictograph to show the data in the table. She used a key of 3. How many pictures did she draw for cats?

FAVORITE PETS	
Dog	18
Cat	24
Bird	12
Fish	6

4 Use a key of 6. Draw a new pictograph to show the information in the table.

5 Sam made a pictograph using the data in the table. He drew 3 pictures for fish. What key did he use?

6 Using a key of 2, how many pictures did Brandon draw for dogs?

7 How many pictures did Brandon draw for birds?

▶ **Quiz** ....

Make a graph for each problem.

8 Sara grew 14 flowers, Evan grew 16, and Fred grew 10. Make a pictograph to show this information.

9 Alice grew 20 flowers. Using a key of 2, explain how you would show her flowers in a pictograph.

10 Alan, Marcia, Dale, and Regina mowed lawns to earn money. Alan mowed 4 lawns, Marcia mowed 12, Dale mowed 24, and Regina mowed 16. Use a key of 2. Draw a pictograph to show this information.

11 Use a key of 4. Draw a new pictograph for the information in Problem 11.

Problem Solving Skill 17: Estimate or Measure

Alan needs to water 2 plants. Each plant needs about 1 cup of water. He has a 1-pint pitcher to use. About how much should he give each plant? Should Alan estimate or measure the exact amount of water for each plant?

UNDERSTAND

Reread the problem.
- Underline what you are to find.
- Circle the facts.

PLAN

The problem says the plants need *about* one cup. So, you can estimate how much water to give the plants.

SOLVE

2 cups equal one pint. So, Alan can pour about half the pitcher into each plant.

CHECK

- Alan is making punch. The directions say to use 1 cup of punch mix with 1 quart of water. Should he estimate or measure? Explain.

▶ Try These

Tell if you need to measure or if an estimate will do.

1. Lori gives her dog about 1 quart of water every day.

2. Kevin needs 2 cups of water for every 1 cup of pasta.

3. Julian feeds his fish about one teaspoon of food.

4. Karl uses about 1 tablespoon of chocolate in his cup of milk.

PRACTICE ON YOUR OWN

**Tell if you need to measure or if an estimate will do.
Then solve.**

For 1–2, use the table.

SUPPLIES	
Item	**Size**
Yarn	10 ft roll
Yarn	20 ft roll
Macaroni	6 oz bag
Macaroni	1 lb bag

1 James is making 2 pictures with macaroni. He needs 8 ounces of macaroni for each picture. Which package should he buy? Explain.

2 Ashley is making 10 bracelets for her friends. For each bracelet she needs 11 inches of yarn. Which roll of yarn should she buy? Explain.

3 Kenneth has 1 gallon of milk. He pours about 1 quart into a pitcher. How much does he have left?

4 Val has 2 feet of yarn. She wants to make 3 necklaces of the same length. The first piece she cut was 8 inches long. Does she have enough left to make two more necklaces?

5 Mei is putting ribbon on a poster. She needs four 11-inch pieces. About how many feet of ribbon should she buy?

▶ Quiz

**Tell if you need to measure or if an estimate will do.
Then solve.**

6 Roy needs an 8-ft piece of cloth to make a banner. Cloth is only sold in yards. About how many yards of cloth will Roy need?

7 Pat pours about 4 cups of tea out of a 2-quart pitcher. How much tea is left in the pitcher?

Problem Solving Strategy 18: Make a Table

Each student in art class needs 100 centimeters of yarn. There are 5 students in the art class. How many meters of yarn are needed?

UNDERSTAND

Reread the problem.
- Underline what you are to find.
- Circle the facts.

PLAN

Think about the problem and decide which strategy to use. You know that you need to change centimeters to meters. So, you can *make a table* to see how much yarn the art class needs.

SOLVE

You can make a table to show the relationship between the number of centimeters and meters.

Remember: 100 cm = 1m

Students	1	2	3	4	5
Centimeters	100	200	300	400	500
Meters	1	2	3	4	5

_5_ students need _500_ cm of yarn.
_500_ cm = _5_ m

So, _5_ m of yarn are needed.

CHECK

- How can you check your answer?

▶ Try These

Use *make a table* to solve.

1. Use the table to decide how many meters of yarn are needed for 8 students.

2. How many meters of yarn are needed for 6 students?

Name ____

PRACTICE ON YOUR OWN

Use *make a table* to solve.

1 Kaleb needs 200 centimeters of blue yarn and 100 centimeters of red yarn. How many meters of yarn does he need in all?

2 Sandra has 2 meters of green yarn and 2 meters of yellow yarn. How many centimeters of yarn does she have in all?

For 3–8, use the tables.

3 George rolled a marble 2 meters. How many decimeters did the marble roll?

Centimeters	10	20	30
Decimeters	1	2	3

Decimeters	10	20	30
Meters	1	2	3

4 Marcus jumped 2 meters. How many centimeters did he jump?

Centimeters	100	200	300
Meters	1	2	3

5 Lucy's hand is 1 decimeter long. How many centimeters long is her hand?

6 A fork is about 2 decimeters long. About how many centimeters long is it?

7 Mrs. Green's desk is 2 meters long. How many decimeters long is her desk?

8 The height of Stan's room is 300 centimeters. How many meters high is it?

▶ Quiz

Use *make a table* to solve.

9 Benny needs 60 decimeters of lace to make a gift. How many meters of lace does he need in all?

10 Jo needs 2 meters of red yarn and 1 meter of blue yarn. How many centimeters of yarn does she need in all?

11 Sue has 20 centimeters of purple ribbon and 40 centimeters of orange ribbon. How many decimeters of ribbon does she have in all?

12 Ty has a piece of rope that is about 7 meters long. About how many decimeters long is Ty's rope?

Problem Solving Strategy 19: Draw a Diagram

Mr. Hall drew these plane figures. How are the figures alike, and how are they different?

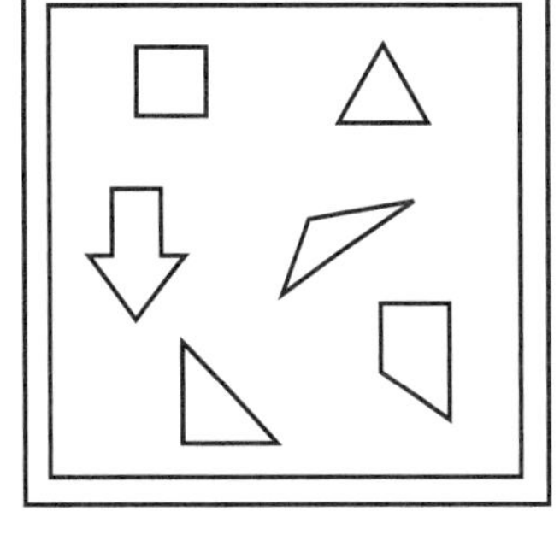

UNDERSTAND

Reread the problem.

- Underline what you are to find.
- What information will you use?

PLAN

You can use the strategy *draw a diagram* to solve the problem.

SOLVE

A **Venn diagram** shows relationships among groups. Put each figure in the oval that describes it. If both ovals describe the figure, put it in the space where the ovals overlap.

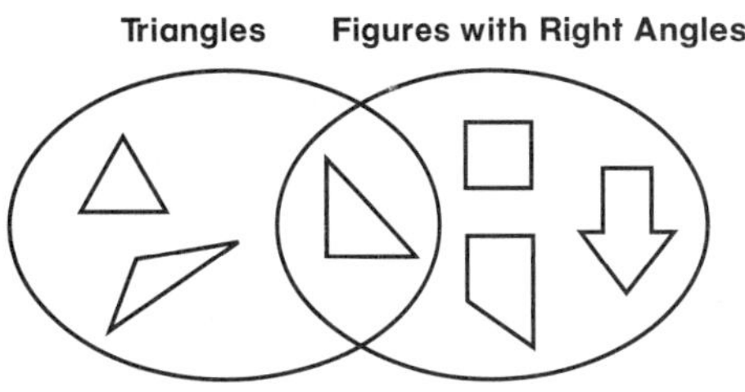

Into what two groups can you separate the figures?

Do any figures fit in both groups?
Where can the figure be placed?

CHECK

- How do you know the right triangle fits in both groups?

▶ Try These

1. Where would the figure at the right go in the Venn diagram above? 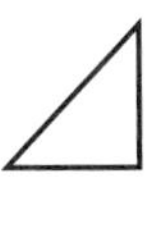

2. Where would a rectangle go in the Venn diagram above?

PRACTICE ON YOUR OWN

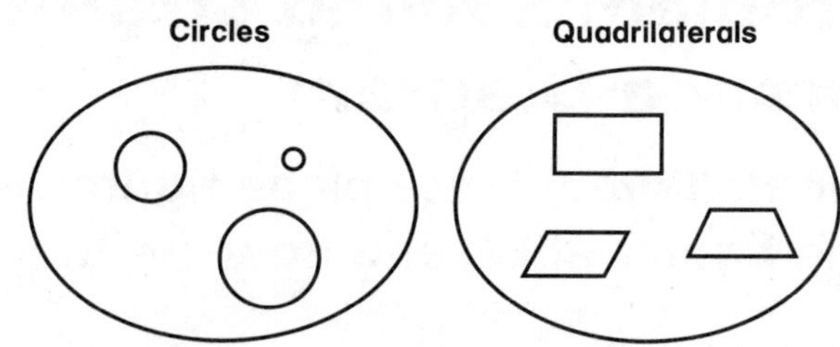

1 Some Venn diagrams show sets that do not overlap. Explain why the sets to the right do not overlap.

2 Cassie used the labels *multiples of 5* and *even numbers* in her Venn Diagram. List 3 numbers that would go in the space that overlaps.

3 In Cassie's Venn diagram, where would the number 32 be placed?

4 Where would the number 55 go in Cassie's Venn diagram?

5 Is the number 13 in Cassie's Venn diagram? Explain.

6 Max used the labels *multiples of 3* and *even numbers*. List 3 numbers that would go in the space that overlaps.

7 Where would the numbers 3, 21, and 27 go in Max's Venn diagram?

▶ Quiz

8 Devon made a Venn diagram. He labeled the ovals *multiples of 7* and *odd numbers*. List 3 numbers that would go in the space that overlaps.

9 Where would the number 14 go in Devon's Venn diagram?

10 Make a Venn diagram for the numbers at the right. Which of these numbers go in the space that overlaps?

Multiples of 3: 6, 9, 12, 15, 18, 21
Multiples of 4: 8, 12, 16, 20, 24

Problem Solving Strategy 20: Make a Model

Petra made a big triangle with pattern blocks. She used 4 small triangles. What is another combination of pattern blocks she can use to make a triangle congruent to the one she made?

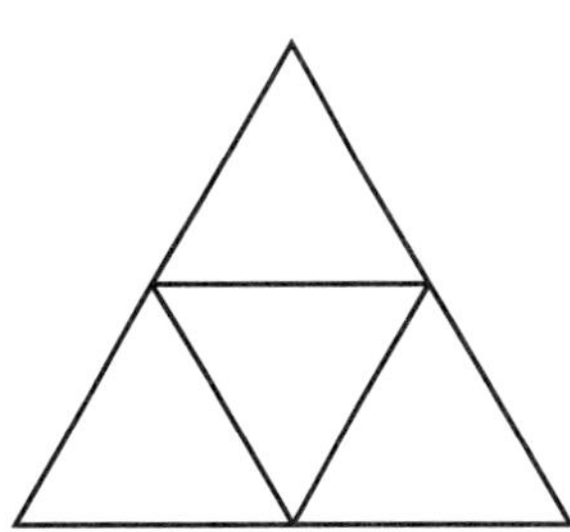

UNDERSTAND

Reread the problem.
- Underline what you are to find.
- Circle the facts.

PLAN

You can use the strategy *make a model* to solve the problem.

SOLVE

How can you use the strategy to solve the problem? How can you exchange some triangles for a different pattern block?

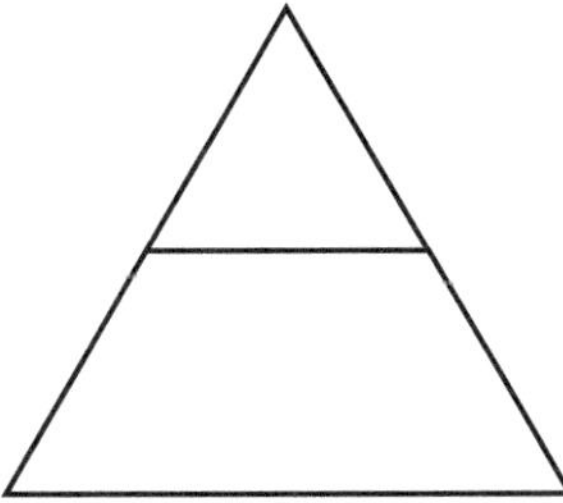

So, Petra can use ____1 trapezoid____ and ____1 triangle____ to make a congruent triangle.

CHECK

- How can you check that the triangles are congruent?

▶ Try These

Make a model to solve.

1. What is another way to model Petra's triangle?

2. How many triangles make up a rhombus?

Name _______________________________________

PRACTICE ON YOUR OWN

Make a model to solve.

1. Which two pattern blocks make a figure that is congruent to the trapezoid?

2. How many triangles make a hexagon?

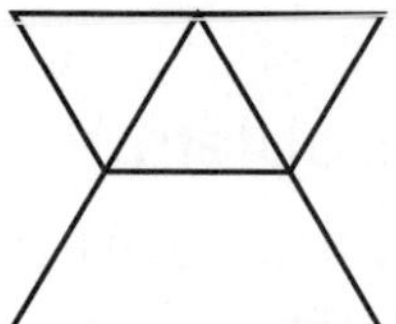

3. Jack made the figure to the right. What is another combination of pattern blocks that can make a congruent figure?

4. Use a different combination of pattern blocks to make another figure congruent to the one that Jack made.

5. How many rhombuses make a hexagon?

6. Tina made the hexagon to the right. What is another combination of pattern blocks that can make a congruent hexagon?

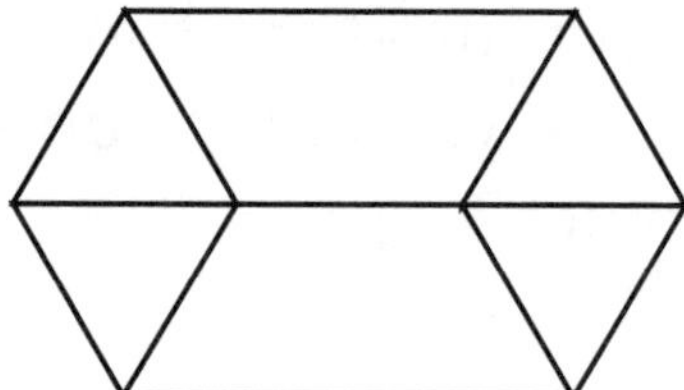

▶ Quiz

Make a model to solve.

7. Rob made the big rhombus to the right. How many triangles do you need to make a congruent rhombus using only triangles?

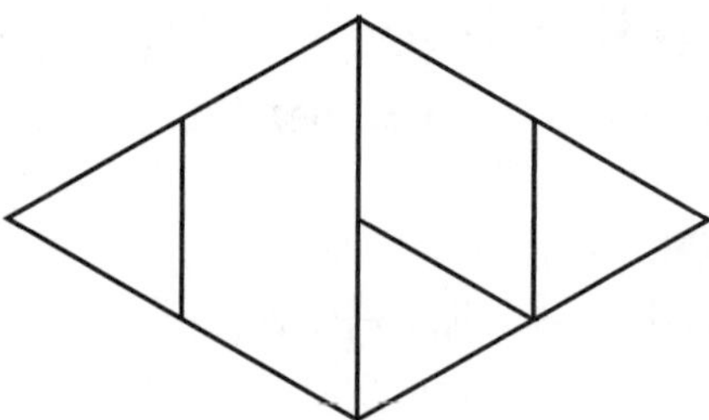

8. Gina only has trapezoids and rhombuses. How can she make a rhombus that is congruent to Rob's rhombus?

9. Use a different combination of pattern blocks to make another rhombus congruent to Rob's rhombus.

Problem Solving Skill 21: Identify Relationships

Mrs. Green drew this plane figure. It is the side view of some solid figures. Which solid figures could this be?

UNDERSTAND

Reread the problem.

- Underline what you are to find.
- Circle the facts.

PLAN

Think about the solid figures that you know. Which of these figures look like triangles when you look at them from the side?

SOLVE

List the solid figures that look like triangles from the side.

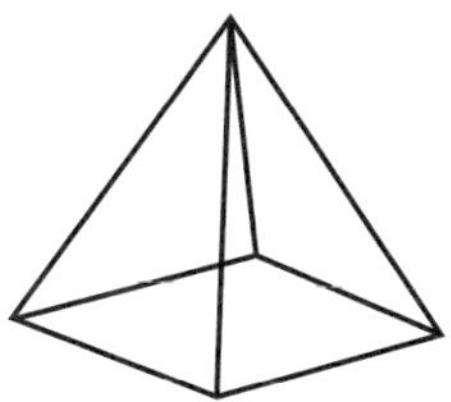

Square Pyramid

Cone

The face of a square pyramid is a triangle.

The side of a cone looks like a triangle.

So, a triangle could be the side of a ___square pyramid___ or a ___cone___ .

CHECK

- Which solid figure looks like a circle from all directions?

▶ Try These

1 What plane figure forms a face of a cube?

2 What plane figure forms the bottom of a cone?

PRACTICE ON YOUR OWN

1 Imran drew a rectangular prism. What plane figure did he use to draw the side?

2 Name two solid figures that have a square base.

3 Julie drew the side view of a cylinder. What plane figure did she use?

4 What plane figure can show the side view of a cone?

5 Dale drew a circle to show the top view of his soda can. What solid figure describes the shape of a soda can?

6 Matt drew a solid figure using only triangles and one square. What solid figure did he draw?

7 Which of these solid figures has the same face on every side: cylinder, rectangular prism, or cube?

8 Nina drew the side view of a square pyramid. What plane figure did she draw?

9 Lori made a solid figure out of construction paper and tape. She used two congruent circles and a rectangle. What solid figure did she make?

10 Heidi drew a solid figure using only triangles. What solid figure did she draw?

11 Which solid figure has more faces, a square pyramid or a triangular pyramid? How many more?

▶ Quiz

12 Tracy drew a triangle to show the side view of two solid figures. Which two solid figures could the triangle show?

13 Rob drew a rectangle to show the side view of a skyscraper. Which two solid figures could describe the skyscraper?

14 Lucia wants to draw a picture of her baseball. What plane figure should she use?

15 Bo wants to draw the side view of a cube. What plane figure should he use?

Problem Solving Skill 22:
Make Generalizations

Kim has 16 feet of fence to put around her garden. She can make a square or a rectangle. If she wants the most area possible, should she make a square or a rectangle?

UNDERSTAND

Reread the problem.
• Underline what you are to find.
• Circle the facts.

PLAN

You can draw pictures that show the square and rectangular gardens she can make.

SOLVE

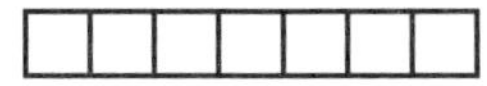 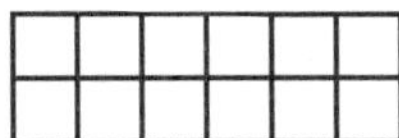 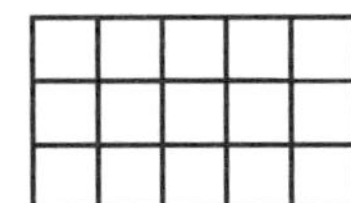 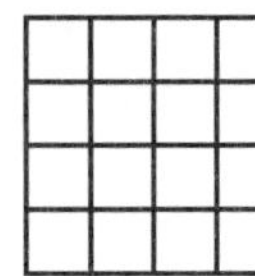

Perimeter:
$7 + 1 + 7 + 1 = 16$ ft
Area:
$7 \times 1 = 7$ sq ft

Perimeter:
$6 + 2 + 6 + 2 = 16$ ft
Area:
$6 \times 2 = 12$ sq ft

Perimeter:
$5 + 3 + 5 + 3 = 16$ ft
Area:
$5 \times 3 = 15$ sq ft

Perimeter:
$4 + 4 + 4 + 4 = 16$ ft
Area:
$4 \times 4 = 16$ sq ft

Order the areas. 16 sq ft $>$ 15 sq ft $>$ 12 sq ft $>$ 7 sq ft
16 square feet is the greatest area.
So, Kim should make a ____square____ garden.

CHECK

• How did you find the areas of the square and rectangle?

▶ Try These

Solve.

1. What if Kim had 12 feet of fencing to make a rectangle with the greatest possible area? How long should it be? How wide should it be?

2. What if Kim wanted to make a square with her 12 feet of fencing? How long would it be? How wide?

PRACTICE ON YOUR OWN

Solve.

1. Abby made a square box for her kitten with 4 feet of wood. How long was each side of the box?

2. What was the area of Abby's box?

3. Chuck used 10 feet of wood to make a rectangular box for his bunny. What was the length of the box? What was the width?

4. What was the area of Chuck's box?

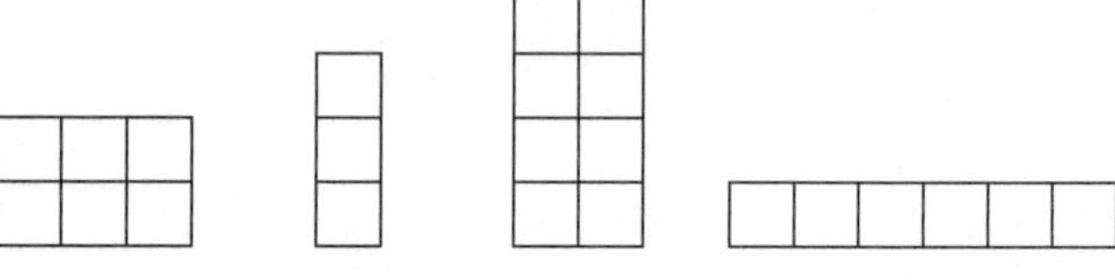

5. Wendy drew some figures on grid paper. Which figure has an area of 8 square units?

6. Which of Wendy's figures has a perimeter of 8 units?

7. Which of Wendy's figures has the greatest perimeter?

▶ Quiz

Solve.

8. Bill wants to make a rectangular garden with 16 feet of fencing. He wants the area to be 12 square feet. What is the length of the garden? the width?

9. Paula has 28 feet of wood to build a fort. She can build a square or a rectangle. If she wants the most area possible, should she build a square or a rectangle?

10. Jesse used 36 feet of fencing to make a square play space for his rabbit. What was the length of each side? What was the area?

11. How could Jesse use the 36 feet of fencing to make a rectangular play space with an area of 72 feet?

Problem Solving Strategy 23:
Find a Pattern

Ann needs to deliver flowers to 165 Beach Ave. Some of the house numbers are hidden. To which house should she deliver the flowers? Use the map for help.

UNDERSTAND

Reread the problem.
- Underline what you are to find.
- Circle the facts in the problem and on the map.

PLAN

What strategy can you use to solve the problem?

SOLVE

Step 1
Find the rule. Use the house numbers you can see.

$$+5 \qquad +5$$
$$140 \rightarrow 145 \rightarrow 150 \rightarrow \underline{\quad} \rightarrow 160 \rightarrow \underline{\quad}$$
Rule: the house numbers increase by 5.

Step 2
Use the rule to find the missing house numbers.

$$150 + 5 = 155 \qquad 160 + 5 = 165$$

The missing house numbers are 155 and 165.

So, Ann should deliver flowers to the ___6___ th house on Beach Ave.

CHECK

- How can you use subtraction to check your answer?

▶ Try These

Use *find a pattern* to solve.

1. What would be the numbers of the next 3 houses after 165 Beach Ave?

2. Melinda made this pattern. What numbers are missing?
   81, 84, 87, ____, ____, 96

PRACTICE ON YOUR OWN

Use *find a pattern* to solve.

1 What is the rule for Mary's pattern?
35, 31, 27, ___, 19, ___

2 What are the missing numbers in Mary's pattern?

3 Mrs. Hale gave this pattern to her class. What is the rule?
132, 134, 136, 138, 140

4 What are the next three numbers in Mrs. Hale's pattern?

5 Phil wrote this pattern.
77, 72, 67, ___, 57, ___
What is the rule?

6 What are the missing numbers in Phil's pattern?

7 Laura wrote these numbers.
89, 87, 85, ___, ___, 79
What is the rule?

8 What are the missing numbers in Laura's pattern?

9 What are the first three numbers in Jake's pattern?
___, ___, ___, 50, 53, 56

10 What are the next three numbers in Jake's pattern?

▶ Quiz

Use *find a pattern* to solve.

11 What is the rule for Amy's pattern?
154, 158, 162, ___, 170, ___

12 What are the missing numbers in Amy's pattern?

13 Gene wrote these numbers.
19, 25, 31, 37, 43, 49
What is the rule?

14 What are the next three numbers in Gene's pattern?

15 What are the first three numbers in Matt's pattern?
___, ___, ___, 253, 250, 247

16 What are the next three numbers in Matt's pattern?

Problem Solving Strategy 24:
Make a List

**Mr. Sanchez is planning games for his class.
The games are soccer, baseball, and kickball.
How many ways can he order these games?**

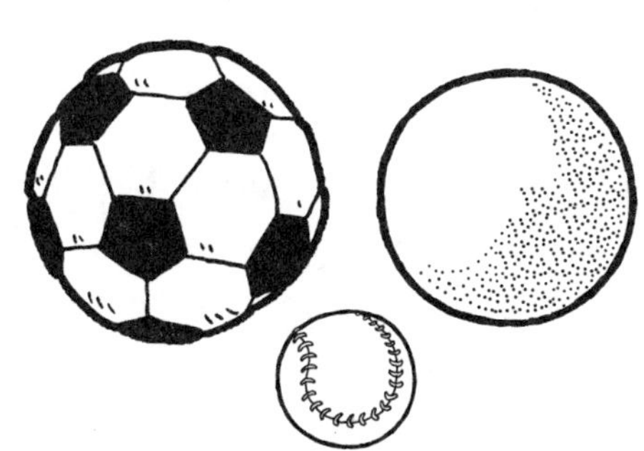

UNDERSTAND

Reread the problem.

- Underline what you are to find.
- Circle the facts.

PLAN

Since you have to find the different ways to order the games,
you can *make a list* to keep track of the different combinations.

SOLVE

- Start your list by choosing a
  game to be Game 1.
- Choose a different game to
  be Game 2.
- Choose the remaining game
  to be Game 3.

Game 1	Game 2	Game 3
soccer	baseball	kickball
soccer	kickball	baseball
baseball	soccer	kickball
baseball	kickball	soccer
kickball	soccer	baseball
kickball	baseball	soccer

Now, count the number of different ways the games can be ordered.
So, there are __6__ different ways to order the games.

CHECK

- What other strategy could you use to solve this problem?

▶ Try These

Make a list to show the number of possible combinations.

1 How many ways can you place
the numbers 1, 2, and 3 in a
3-digit number?

2 How many ways can you order
the letters A and M?

Name ____

PRACTICE ON YOUR OWN

Make a list to show the number of possible combinations.

1. How many even 3-digit numbers can you make with the numbers 4, 5, and 2?

2. How many ways can Bob and Sam stand in line?

3. List the different ways that Ray, Mae, and Jay can stand in line. In how many ways is Jay first in line?

4. Kay is planting daisies, roses, tulips, and pansies. In how many ways can she plant these flowers in a row?

5. List the ways you can put 1, 8, 6, in a 4-digit number.

6. Lacy layers ice cream, cake, and syrup to make dessert. List the ways he can layer the ingredients.

▶ Quiz

Make a list to show the number of possible combinations.

7. Deon is putting his toys on a shelf. He has a car, a ball, and a bear. How many ways can he place his toys on the shelf? In how many ways is the bear placed first?

8. List the ways you can place the numbers 7, 1, and 5 in a 3-digit number. What is the largest number you can make with these numbers?

9. How many odd 3-digit numbers can you make with the numbers 1, 6, and 8?

10. How many even 2-digit numbers can Harry make with 2 and 3?

Problem Solving Strategy 25:
Make a Model

Two students ran a relay race. Pete ran $\frac{3}{4}$ mile and Lisa ran $\frac{1}{2}$ mile. Who ran farther?

UNDERSTAND

Reread the problem.
- Underline what you are to find.
- Circle the facts.

PLAN

Make a model to see which fraction is greater.

SOLVE

Model with fraction bars the distance that each student ran.

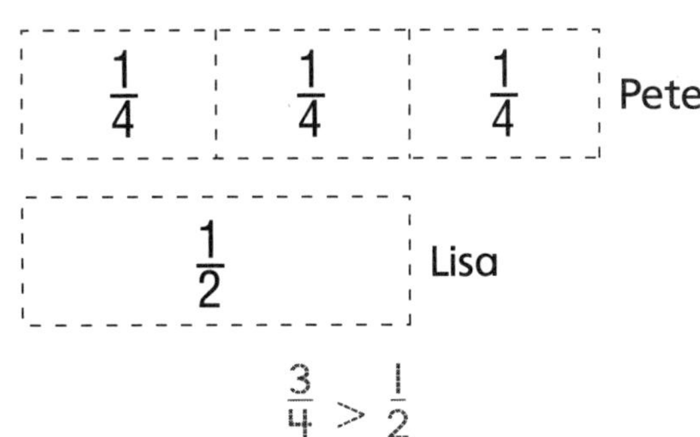

Step 1
Line up the fraction bars for $\frac{3}{4}$ and $\frac{1}{2}$.

Step 2
Compare the lengths of the fraction bars.

Step 3
Decide which set of fraction bars is longer.

$$\frac{3}{4} > \frac{1}{2}$$

So, Pete, who ran $\frac{3}{4}$ of a mile, ran farther.

CHECK

- If Lisa had run $\frac{7}{8}$ of a mile, who would have run the farthest? Explain.

▶ Try These

Use *make a model* to solve.

1. In a soccer game, Bill played $\frac{3}{5}$ of the game and Kyle played $\frac{7}{8}$ of the game. Who played longer?

2. In another race, Tina ran $\frac{5}{6}$ of a mile and Jon ran $\frac{2}{3}$ of a mile. Who ran farther?

Name _______________________________

PRACTICE ON YOUR OWN

Use *make a model* to solve.

1. Larry, Jane, and Owen shared a pizza. Larry ate $\frac{1}{4}$ of the pizza, Jane ate $\frac{3}{8}$, and Owen ate $\frac{1}{6}$. Who ate the most?

2. Shane mixed $\frac{3}{4}$ cup of ice cream and $\frac{5}{8}$ cup of milk to make a shake. Of which ingredient did he use more?

3. Sandra has finished $\frac{1}{2}$ of her science project. Megan has finished $\frac{2}{5}$ of her science project. Who has finished more?

4. Guy collects toy trucks. $\frac{2}{3}$ of his trucks are yellow and $\frac{2}{6}$ of his trucks are red. Of which color does Guy have more?

5. Tomas will spend $\frac{1}{4}$ of his Saturday reading, $\frac{4}{10}$ of his day playing outside, and $\frac{1}{8}$ of his day drawing. On which activity will he spend the most time?

6. Janna has visited $\frac{1}{3}$ of the states in the United States. Nate has visited $\frac{4}{10}$ of the states. Who has visited fewer states?

▶ Quiz

Use *make a model* to solve.

7. Sam played $\frac{6}{8}$ of a basketball game on Saturday. He played $\frac{5}{6}$ of a basketball game on Sunday. On which day did he play longer?

8. Maria and Omar are reading the same book. Maria has finished $\frac{2}{3}$ of the book. Omar has finished $\frac{3}{5}$ of the book. Who has read more?

9. Katie and Nicholas share a school project. Katie has finished $\frac{1}{3}$ of her part. Nicholas has finished $\frac{4}{6}$ of his. Who has more to go?

10. Eddie followed a recipe. It called for $\frac{1}{2}$ teaspoon of salt and $\frac{3}{5}$ teaspoon of pepper. Of which ingredient did he use less?

Problem Solving Skill 26: Reasonable Answers

Cara's mother made a pie. Cara and Mary each ate $\frac{1}{4}$ of the pie. What fraction of the pie was left?

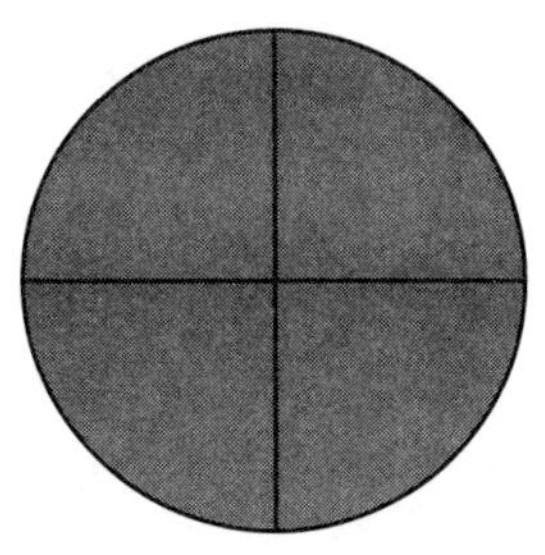

UNDERSTAND

Reread the problem.
- Underline what you are to find.
- Circle the facts.

PLAN

You know that there are $\frac{4}{4}$ in one whole. So, if Cara and Mary each ate $\frac{1}{4}$, you can subtract $\frac{2}{4}$ from the whole to see how much of the pie is left.

SOLVE

Start with the whole pie, or $\frac{4}{4}$.

Take away $\frac{1}{4}$ for Cara.

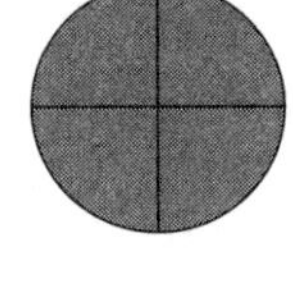

Take away another $\frac{1}{4}$ for Mary.

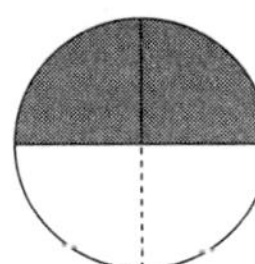

So, ___$\frac{2}{4}$___ , or ___$\frac{1}{2}$___ , of the pie is left.

CHECK

- How do you know that your answer is reasonable?

▶ Try These

Solve. Tell how you know your answer is reasonable.

1. Cara's mother ate another $\frac{1}{4}$ of the pie. Now how much of the pie is left?

2. Ann ran $\frac{2}{6}$ of the way home and rested. Then she ran $\frac{3}{6}$ of the way home. How much farther does she have to go?

PRACTICE ON YOUR OWN

Solve. Tell how you know your answer is reasonable.

1. Paul planted tulips. He planted $\frac{2}{5}$ of the tulips on Monday and the rest on Tuesday. What fraction of the tulips did he plant on Tuesday?

2. Julie made a can of soup. She ate $\frac{1}{3}$ of the can and left the rest. How much soup is left?

3. Henry finished $\frac{3}{8}$ of his homework before dinner. He finished the rest after dinner. How much of his homework did he finish after dinner?

4. Louie read $\frac{1}{2}$ of his book on Saturday and $\frac{1}{2}$ of his book on Sunday. How much of his book does he have left to read?

5. Of all the students in Mr. Tale's class, $\frac{3}{5}$ are boys. What fraction of the class is girls?

6. John and Ryan each ate $\frac{3}{8}$ of a pizza. How much of the pizza is left?

▶ **Quiz**

Solve. Tell how you know your answer is reasonable.

7. Fran has listened to $\frac{1}{3}$ of the songs on a CD. How much of the CD is left to listen to?

8. Nick's dog ate $\frac{4}{5}$ of the bag of dog food. How much of the bag of dog food is left?

9. Your aunt gives you a bag of raisins to share with your friend. You eat $\frac{1}{7}$ of the bag. Your friend eats $\frac{2}{7}$ of the bag. What fraction of the raisins is left?

10. At a family reunion, the girls ate $\frac{3}{8}$ of the watermelon. The boys ate $\frac{5}{8}$ of the watermelon. How much of the watermelon was left?

Problem Solving Skill 27:
Too Much/Too Little Information

Mr. Fay bought 1.5 pounds of lemons, 3 pounds of peaches, 2.5 pounds of berries, and 1 pound of sugar. List the weights of the fruit in order from greatest to least.

UNDERSTAND

Reread the problem.
- Underline what you are to find.
- Circle the facts.

PLAN

Find the facts you need to solve the problem. Cross out any information you do not need.

SOLVE

Find the weights of the fruit.

1.5 pounds of lemons
3 pounds of peaches
2.5 pounds of berries

What items are not fruit?

1 pound of sugar

Order the weights of the fruits.

$3 > 2.5 > 1.5$

So, the order is __3__ pounds of peaches, __2.5__ pounds of berries, and __1.5__ pounds of lemons.

CHECK

- What if the problem asked you to order the weights of all the items that Mr. Fay bought?

▶ Try These

**Write *too much*, *too little*, or *right amount* for each problem.
Solve those with too much or the right amount of information.
Tell what is missing for those with too little information.**

1 Rob ate 4 slices of pizza and Liz ate 3. How many slices are left?

2 Jane bought 2 pizzas. Each pizza had 8 slices. How many slices did she buy in all?

Name ____

PRACTICE ON YOUR OWN

**Write *too much, too little,* or *right amount* for each problem.
Solve those with too much or the right amount of information.
Tell what is missing for those with too little information.**

1. Mrs. Tomas made cupcakes. She gave 5 cupcakes to Joe and his friends. How many cupcakes are left?

2. Mr. Lopez bought 8 paper plates and 10 paper cups. He gave 5 plates to Jose and his friends. What fraction of plates are left?

3. Mrs. Greene made 8 goody bags. She gave 5 to Jess and her friends. How many goody bags are left?

4. Harry bought two hats and two scarves. Each hat cost $15. He gave the cashier $40. What was his change?

5. Jim bought sunglasses and three shirts. Each shirt cost $10. The sunglasses cost $8. How much did he spend on shirts?

6. At practice Gina ran 3.1 miles, Ann ran 2.8 miles, and Suzy ran 1.9 miles. Who ran the farthest?

▶ **Quiz**

**Write *too much, too little,* or *right amount* for each problem.
Solve those with too much or the right amount of information.
Tell what is missing for those with too little information.**

7. Sylvia made 4 necklaces and 3 bracelets. She sold each necklace for $3. She sold each bracelet for $2. How much will she make if she sells them all?

8. Kerry, Sam, and Paul all walk to school. Kerry walks 2.1 miles. Sam walks 1.5 miles. How many miles do the three walk altogether?

9. Dino has 25 minutes of math homework, 20 minutes of spelling homework, and 30 minutes of piano lessons tonight. How many minutes will he spend on schoolwork?

10. Nico bought 3 pizzas. Each pizza cost $9.98 and had 12 slices. How many slices of pizza did Nico buy in all?

Problem Solving Strategy 28: Solve a Simpler Problem

Sue has $10 to buy fruit. If she buys 1 pound of apples and 2 pounds of mangos, how much money will she have left?

UNDERSTAND

Reread the problem.

- Underline what you are to find.
- Circle the facts.

PLAN

You can *solve a simpler problem* to find the answer.

SOLVE

Step 1:
Find the price for each fruit.

1 lb of apples: $1.00
2 lb of mangos: $2.00 + $2.00

Step 2:
Add to find the total cost.

$1.00 apples
+ $4.00 mangos
$5.00

Step 3:
Subtract the total cost of the fruit from $10.

$10.00 money Sue has
− $ 5.00 total cost of fruit
$ 5.00 Sue's change

So, Sue will have __$5.00__ left.

CHECK

- Is your answer reasonable? Explain.

▶ Try These

Use the prices above. *Solve a simpler problem* to solve.

1. Sue buys 2 pounds of pears and 2 pounds of mangos. How much change will she receive from $10?

2. Don has $5. He bought 1 pound of apples and 2 pounds of pears. How much change will he receive?

Name _______________________________

PRACTICE ON YOUR OWN

Solve a simpler problem to solve.

For 1–4, use the prices on page IPS55.

1 Marcia has $7. Does she have enough to buy 2 pounds of pears, 1 pound of mangos, and 1 pound of apples? Explain.

2 Mark has $4. Does he have enough to buy 2 pounds of mangos and 1 pound of apples? Explain.

3 Laura bought 4 pounds of pears and 1 pound of apples. She gave the cashier $10. How much change did she receive?

4 Stu bought 3 pounds of mangos and 1 pound of apples. He gave the cashier $20. How much change did he receive?

5 Alex bought 2 shirts for $8 each. He bought pants for $15. He gave the cashier $40. How much change did he receive?

6 Sheri is 2 years older than Jim. Jim's age is 2 times as great as Lou's age. Lou is 4 years old. How old are Sheri and Jim?

▶ Quiz

Solve a simpler problem to solve.

7 Cindy bought 2 dolls for $5 each. She bought 3 candy bars for $1 each. She gave the cashier $20. How much change did she receive?

8 Ali is 1 year younger than Mae. Mae's age is 3 times as great as Kim's age. Kim is 3 years old. How old are Ali and Mae?

9 Eli bought 3 toy cars for $4 each. He bought 1 toy truck for $6. He gave the cashier $20. How much change did he receive?

10 Sam has $15. Does he have enough to buy 4 baseballs for $2 each and 2 bats for $5 each? Explain.

11 Robin is 5 years older than Leon. Leon's age is 1.5 times as great as Eliza. Eliza is 8 years old. How old are Robin and Leon?

12 Jackson had $25. He paid $5 for breakfast, $7 for lunch, and $12 for dinner. How much did he have left?

Problem Solving Skill 29: Choose the Operation

Maria will work 20 hours each week to get ready for the state fair. The fair is in 3 weeks. How many hours will Maria work?

UNDERSTAND

- Underline what you are to find.
- Circle the facts.

PLAN

Think about the problem and decide which operation to use. Since you are joining equal groups, you multiply.

SOLVE

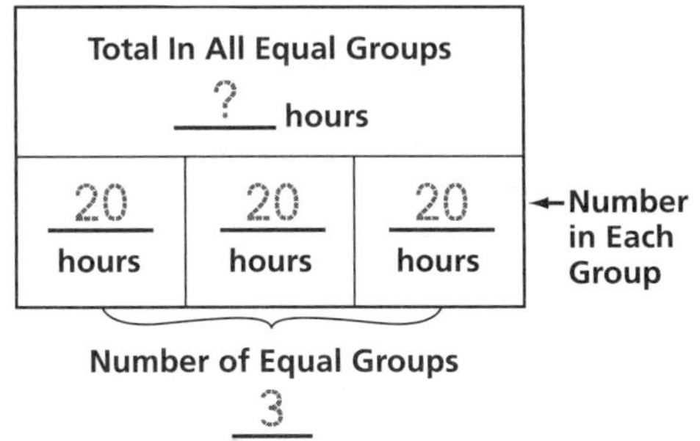

3 groups of 20 hours = total number of hours

$3 \times 20 =$ ___60___

So, Maria will work ___60___ hours.

CHECK

- How do you know that you chose the correct operation?

▶ Try These

Write whether you would *add, subtract, multiply,* or *divide.* Then solve.

1. Carol bought a CD for $14.99. She paid for it with a $20 bill. How much change did she get?

2. Hank bought 64 ounces of orange juice. How many people can get an 8-ounce serving?

PRACTICE ON YOUR OWN

**Write whether you would *add, subtract, multiply,* or *divide*.
Then solve.**

1 Amy paid $12.99 for a T-shirt and $8.99 for a cap. How much more did the T-shirt cost than the cap?

2 Jake paid $5.99 for a baseball cap and $2.00 for a hot dog. How much did he spend in all?

3 Matt cares for his pet rabbits for 30 minutes each day. How many minutes does he spend caring for his pets in 7 days?

4 A total of 36 people will attend a picnic. How many cookies should Mr. Lee bake so that each person can have 2 cookies?

5 The 32 students were put into 4 equal groups. How many were put in each of the 4 groups?

6 Mary baby-sits. She earned $5.80 on Friday and $12.00 on Sunday. How much did she earn in all?

7 Beth makes and sells beaded necklaces. Each one sells for $6. She earned $42 at the fair, how many necklaces did she sell?

8 Jesse earns $12 each time he mows a lawn. Last week he mowed 6 lawns. How much did he earn in all?

9 Ana has a total of 89 stamps. She places 40 stamps in an album. How many are left?

10 Sam bought milk for $2.99, bread for $1.29, and apples for $1.12. How much did he spend in all?

▶ Quiz

**Write whether you would *add, subtract, multiply,* or *divide*.
Then solve.**

11 Mark owes his sister $10.80. He paid her $2.80. How much does he still owe her?

12 Molly bought 6 drinks for $0.75 each. How much did she spend for the drinks?

13 Cindy bought a can of apple juice for $0.95 and a can of orange juice for $0.85. How much did she spend in all?

14 Lin bought 5 packs of baseball cards for a total of $10. How much did each pack of cards cost?

Problem Solving Skill 30:
Interpret the Remainder

Mindy needs 15 cans of juice for the picnic. The juice is sold in packs of 4. How many packs must she buy?

UNDERSTAND

Reread the problem.

- Underline what you are to find.
- Circle the facts.

PLAN

Think about the problem and decide which operation to use. Since you need to know how many groups of 4 there are in 15, you divide.

SOLVE

Find 15 ÷ 4.
If Mindy buys 3 packs of juice,
she will have 12 cans of juice. ⟶
She will still need 3 more cans. ⟶
The remainder says that Mindy needs another pack of juice.

$$\begin{array}{r} 3r3 \\ 4\overline{)15} \\ -12 \\ \hline 3 \end{array}$$

So, Mindy must buy __4__ packs of juice.

CHECK

- How can you multiply and add to check?

▶ Try These

Interpret the remainder to solve.

1 Juan is making brownies for the picnic. He needs 2 cups of brownie mix for each pan of brownies. If he has 11 cups of brownie mix, how many pans can he make?

2 What if Juan had 9 cups of brownie mix? How many pans of brownies could he make?

PRACTICE ON YOUR OWN

Interpret the remainder to solve.

1. What if Mindy had to bring 26 cans of juice? If there are 4 cans of juice in a pack, how many packs would she need to buy?

2. Teresa has 45 stickers in her collection. She can put 7 stickers on a page. How many pages does she need?

3. Hal worked on a math lesson for 21 minutes. Each problem took him 5 minutes to complete. How many problems did he complete?

4. Jill is making bows for gifts. Each bow takes 3 feet of ribbon. She has 22 feet of ribbon. How many bows can she make?

5. Bill read a book for 45 minutes. Each chapter took 7 minutes. How many chapters did he finish?

6. A class of 18 students is going on a field trip. Each car holds 4 students. How many cars does the class need?

7. Mrs. Lane divided her 23 students into groups of 3. How many groups are needed for every student to be included?

8. Adam finished his homework in 35 minutes. If each assignment took 10 minutes, how many assignments did he complete?

▶ **Quiz** ....

Interpret the remainder to solve.

9. Jana needs to buy 39 cans of soda for her party. Soda is sold in packs of 6. How many packs should Jana buy?

10. Francisco has 53 baseball cards. Each page of his album holds 6 cards. How many pages does he need?

11. Hank needs 9 inches of wire to hang a picture. He has 40 inches of wire. How many pictures can he hang?

12. Helen has $59 to spend at the music store. Each CD costs $15. How many CDs can Helen buy?

Problem Bank

1 Cheryl made 36 cookies. 15 were sugar cookies, and the rest were oatmeal. How many cookies were oatmeal?

2 On Monday, Jess read 25 pages of his book. On Tuesday he read 30 more pages. How many pages did he read in all?

3 Jim had 28 peanuts. He ate 15. How many peanuts does he have left?

For 4–5, use the model below.

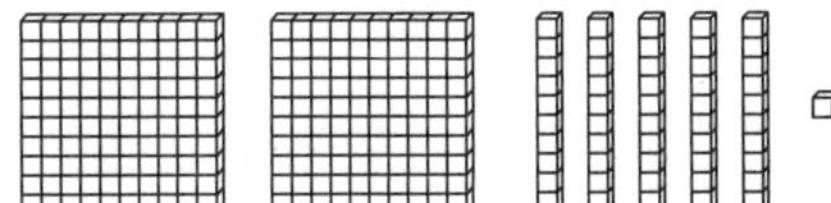

4 Erin made this model with base-ten blocks. What number did she model?

5 What is another way Erin could model the number?

6 Marcus has 1 hundred and 15 tens. How can he model 220?

7 The sum of two numbers is 15. Their difference is 5. What are the two numbers?

8 There are 4 students in line. Sara is before Tom and Jill, but after Liz. Jill is last in line. Who is second?

9 At track practice, Kim and Sam ran 34 laps in all. Kim ran 4 more laps than Sam. How many laps did Kim run?

For 10–12, use the bar graph.

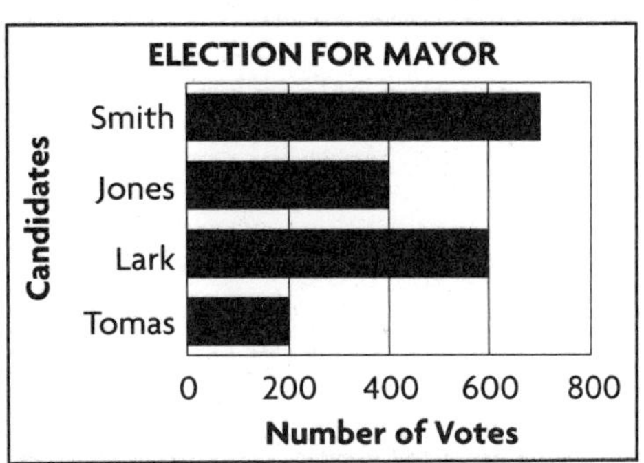

10 If the candidates were listed in order from most votes to least votes, what would be the order of the candidates?

11 Which candidate won the election for Mayor?

12 Which candidate got 400 votes?

13 There are about 500 students at Lincoln school. About 200 students are girls. About how many are boys?

14 Nina has $15. She wants to buy a dress that costs $11.25 and a belt that costs $5. Does she have enough money?

15 Frank has more than 20 baseball cards. He wants to have 50 in all. About how many more cards does he need?

16 Ben has two $1 bills, 2 quarters, 2 dimes, and 5 nickels. How many ways can he pay for a hat that costs $2.25?

Name ____

For 17–18, use the following information. Ann wants to buy a hat that costs $4. She has 8 quarters and three $1 bills.

17 How many ways can she pay for the hat?

18 If Ann wants to keep 4 of her quarters, how will she pay for the hat?

For 19–21, use the time line.

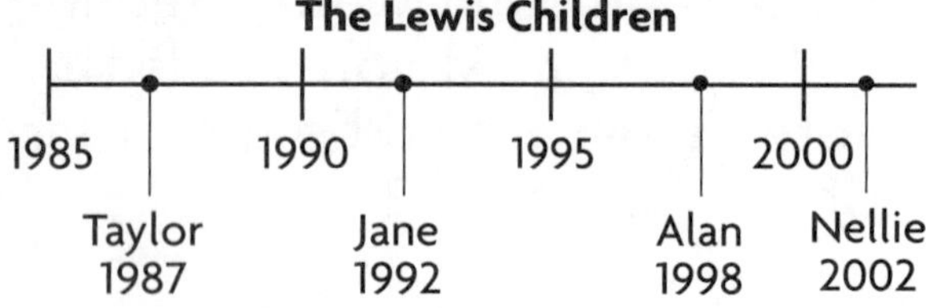

19 Who was born before Nellie and after Jane?

20 Who is the oldest Lewis child, and what year was the child born?

21 Who was born after Alan?

22 Omar bought 10 red pencils and 14 blue pencils. He also bought 8 erasers. How many pencils did he buy in all?

23 Mary bought a coat for $18. She also bought a necklace. She gave the clerk $40. What information do you need to find how much change Mary received?

24 Andy bought 4 boxes of markers. Each box holds 8 markers. One box costs $3. How many markers did he buy?

For 25–27, use the pictograph.

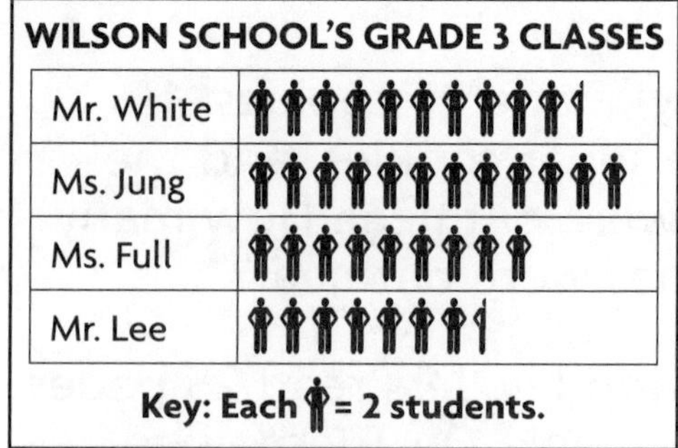

25 Which class has the most students? How many students are there?

26 How many students are in grade 3 at Wilson School?

27 How many more students are in Mr. White's class than in Ms. Full's class?

28 Jordan wrote this pattern on the board: 3, 6, 9, 12. What are the next 4 numbers in Jordan's pattern?

For 29–30, use this pattern: 4, 3, 7, 6, 10, 9, 13, 12, 16.

29 What is the rule for this pattern?

30 What are the next 3 numbers in the pattern above?

31 Gale bought 3 bunches of daisies and 6 roses. There are 10 daisies in each bunch. How many flowers did she buy in all?

32 A child's ticket to the museum costs $2. An adult's ticket costs $4. How much would 2 children's tickets and 2 adults' tickets cost in all?

IPS62 Intervention • Problem Solving

33 Martha has 3 boxes of seashells. Each box holds 9 seashells. If she gives 4 seashells to her friend, how many seashells does she have left?

34 Alicia has 40 stamps in her collection. She puts 5 stamps on each page. Which number sentence below shows how many pages she needs?

$40 + 5 = 45$ $40 \div 5 = 8$ $40 \times 5 = 200$

35 Jane made a hat for each of her 6 friends. She put 9 beads on each hat. Write a number sentence to show how many beads she used in all.

36 Mrs. Adams made 36 cookies. She gave each of her 6 children the same number of cookies. Write a number sentence to show how many cookies each child received.

37 Daniel saw 3 families of deer in the forest. Each family had 4 deer. How many deer did he see in all?

38 Kristin drew 8 pictures of animals and 15 pictures of people. How many more pictures of people did she draw?

39 Pete is watering his flowers. Each flower needs one cup of water. He has a 1-pint pitcher of water. About how much of the pitcher should he use to water each plant?

40 Sherry bought 2 shirts that cost the same amount. She paid with a $20 bill. She received $4 in change. How much did each shirt cost?

41 Josh had 3 boxes of crayons. He lost 5 crayons. He has 25 crayons left. How many crayons were in each box?

42 Lana spent a total of $8 at the fair. She bought a hot dog for $2 and two bags of popcorn. How much did each bag of popcorn cost?

43 Nadine rolled a number cube numbered 1, 2, 3, 3, 3, and 4. She rolled the cube 20 times. Which number do you think she rolled most often?

44 Dennis tossed a coin 30 times. What outcomes are possible? Predict how many times the result will be heads.

For 45–47, use the table.

THIRD GRADERS' FAVORITE DINNERS	
Pizza	20
Salad	15
Pasta	24
Hamburger	8

45 Make a pictograph from the information in the table. Use a key of 2.

46 How many pictures did you list next to pasta?

Name ___________________________

47 How many more pictures are next to pizza than hamburgers?

48 Brad collected 3 white stones, 8 black stones, and 5 brown stones. How many stones did he collect in all?

49 James needs 35 inches of yarn for an art project. About how many feet of ribbon does he need?

50 Trisha needs 5 meters of ribbon for decorations. How many decimeters of ribbon does she need?

51 Patrick needs 250 cm of black yarn and 350 cm of orange yarn. How many meters of yarn does he need in all?

52 Ray needs 400 cm of string for his string-art project. How many meters of string does he need?

For 53, use the figures below.

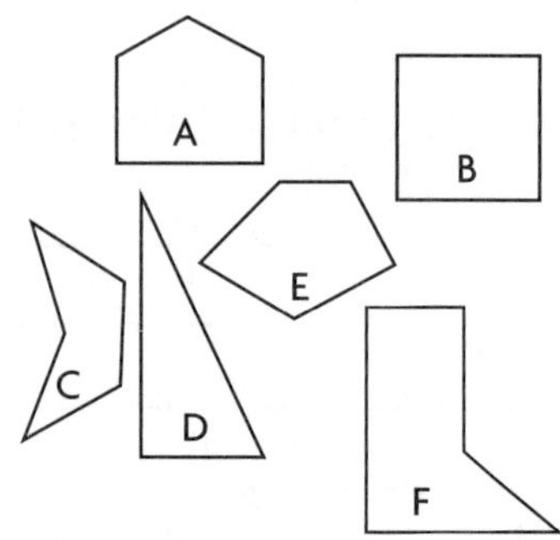

53 Make a Venn diagram to show how the shapes are alike and how they are different. Use the labels *Pentagons* and *Figures with 1 or More Right Angles.*

54 Kelly used the labels *Multiples of 3* and *Multiples of 5* in a Venn diagram. List 2 numbers that could go in the space that overlaps.

55 Randy drew a solid figure using only square faces. Which figure did he draw?

56 Kacey drew a cylinder. Which plane figure did she use to draw the top?

57 Which of these solid figures looks like a circle from the bottom and a triangle from the side: rectangular prism, cylinder, or cone?

For 58–61, use pattern blocks and the drawings below.

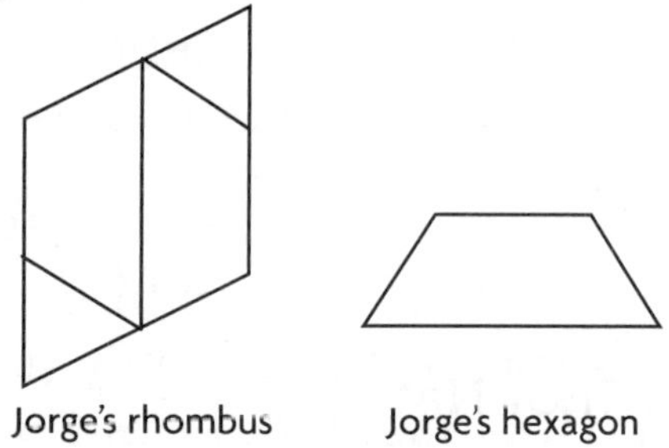

Jorge's rhombus   Jorge's hexagon

58 What other combination of 2 different kinds of pattern blocks can be used to make a rhombus congruent to Jorge's?

59 Four of which pattern blocks can be used to make a rhombus congruent to Jorge's?

60 How many triangle pattern blocks can be used to make a hexagon congruent to Jorge's hexagon?

61 How many trapezoid pattern blocks can be used to make another hexagon congruent to Jorge's hexagon?

62 Al used 16 feet of fencing to build a garden. If he wants the most area possible, should he build a square or a rectangle?

63 Tim wants to make a rectangular garden with 8 feet of fencing. How long will his garden be? How wide will it be?

For 64–66, use the figures below.

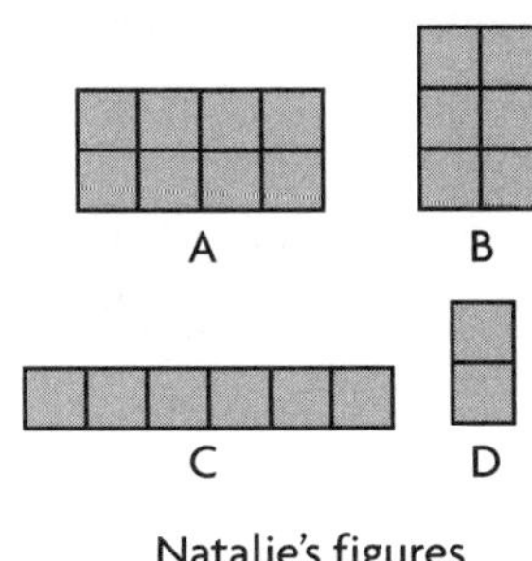

Natalie's figures

64 Which of Natalie's figures has the greatest perimeter?

65 Which of Natalie's figures has the greatest area?

66 Which of Natalie's figures has a perimeter of 10 units?

For 67–68, use this pattern: 354, 361, 368, 375.

67 What is the rule?

68 What are the next four numbers in the pattern?

69 Evan wrote this pattern: 985, 974, 963, ____, 941, ___. What is the rule? What are the missing numbers?

70 Patricia thought of this pattern: 272, 270, ____, 266, ___, 262. What is the rule? What are the missing numbers?

71 Mrs. Wing gave this pattern to her students: ___, ___, ___, 34, 38, 42. What are the first three numbers in the pattern?

72 Marcus eats three different fruits every day. Today he has a kiwi, an apple, and some grapes. In how many ways can he order his eating of a kiwi, an apple, and some grapes?

73 How many ways can you order the colors red, green, yellow, and blue?

74 List all of the 3-digit numbers that you can make from the digits 6, 3, and 9.

75 List the ways that Tara, Kyle, and Fran can stand in line.

76 Mr. Wang and Mrs. Mead have the same number of students in their classes. In Mrs. Mead's class, $\frac{6}{10}$ of the students are boys. In Mr. Wang's class $\frac{4}{5}$ of the students are boys. Which class has more boys?

77 Jules saw $\frac{4}{10}$ of the animals at the zoo. Ann saw $\frac{6}{8}$ of the animals and Andre saw $\frac{3}{5}$ of the animals. Who saw the most animals?

78 Jon and Mike are reading the same book. Jon has read $\frac{1}{3}$ of the book and Mike has read $\frac{1}{2}$. Who has read more?

79 Opal made a trail mix with $\frac{1}{2}$ cup of peanuts, $\frac{1}{4}$ cup of raisins, and $\frac{5}{8}$ cup of chocolate. Of which ingredient did she use the most?

80 Mark ran $\frac{3}{8}$ of a 1-mile race in 2 minutes and another $\frac{3}{8}$ in 1 minute. How much farther does he have to run?

81 Alice and Carrie shared a pumpkin pie. Alice ate $\frac{3}{12}$ of the pie and Carrie ate $\frac{4}{12}$ of the pie. How much is left?

82 Susan and Sharon are writing a report together. Susan has finished $\frac{2}{5}$ of the report and Sharon has finished $\frac{1}{5}$. How much of the report is left?

83 Henry painted $\frac{3}{8}$ of his bedroom on Saturday and $\frac{4}{8}$ on Sunday. How much of his room is left to paint?

84 Courtney has 15 CDs. She sold 10 CDs to the music store. What information do you need to find how much money she made?

85 It takes Alan 45 minutes to walk 3 miles. If he walks 3 miles every day for 6 days, how many miles will he have walked?

86 Tammy made $8.50 on the first day of her garage sale. She made $9 on the second day, and $11.45 on the third day. On which day did she make the most money?

87 Raymond bought two apples and an orange. Each apple cost $0.55. He gave the clerk $3. What information do you need to find how much change he received?

88 Amy bought 3 gumballs for $0.25 each. She bought 2 candy bars for $1 each. She gave the cashier $3. How much change did she receive?

89 Kendra is 3 years older than Will. Will is twice as old as Hanna. Hanna is 4 years old. How old are Kendra and Will?

90 Drew bought 4 golf balls for $2 each. He bought 2 golf clubs for $10 each. He gave the cashier $40. How much change did he receive?

91 Seth is 1 year younger than Juan. Juan is 5 years older than Lyle. Lyle is 4 years old. How old are Seth and Juan?

92 Steve practices piano for 45 minutes each day. How many minutes does he spend practicing piano in 5 days?

93 Sara made $14.78 at her bake sale. Mary made $18.99 at her bake sale. How much more did Mary make than Sara?

94 Kristina bought 4 cakes. Each cake was the same price. She paid $20 for all four cakes. How much was each cake?

95 Pedro bought a skateboard for $10.35 and a helmet for $8.75. How much did he spend in all?

96 James and his 10 friends rode a roller coaster. Each car held 4 people. How many cars did James and his friends use?

97 Zoë needs 45 cans of juice for her party. Juice is sold in packs of 6. How many 6-packs of juice should Zoë buy?

98 Tammy worked on her story for 56 minutes. It took her 8 minutes to write each page. How many pages did she write in all?

99 Alison has 32 photos. She can fit 6 photos on one page in her album. How many pages does she need?

100 Avery and his brother share chores. Every Monday, Wednesday, and Friday Avery washes dishes. He walks the dog every Tuesday, Wednesday, and Thursday. On what day does he do the most chores?